Novelties in Marine Propulsion

Novelties in Marine Propulsion

Guest Editors

Andrea Magrini
Nastia Degiuli
Filippo Avanzi

MDPI Basel • Beijing • Wuhan • Barcelona • Belgrade • Novi Sad • Cluj • Manchester

Guest Editors
Andrea Magrini
Department of Industrial Engineering
Università degli Studi di Padova
Padova
Italy

Nastia Degiuli
Faculty of Mechanical Engineering and Naval Architecture
University of Zagreb
Zagreb
Croatia

Filippo Avanzi
Department of Industrial Engineering
Università degli Studi di Padova
Padova
Italy

Editorial Office
MDPI AG
Grosspeteranlage 5
4052 Basel, Switzerland

This is a reprint of the Special Issue, published open access by the journal *Journal of Marine Science and Engineering* (ISSN 2077-1312), freely accessible at: https://www.mdpi.com/journal/jmse/special_issues/9Q05C89SRB.

For citation purposes, cite each article independently as indicated on the article page online and as indicated below:

Lastname, A.A.; Lastname, B.B. Article Title. *Journal Name* **Year**, *Volume Number*, Page Range.

ISBN 978-3-7258-7881-9 (Hbk)
ISBN 978-3-7258-7882-6 (PDF)
https://doi.org/10.3390/books978-3-7258-7882-6

Contents

About the Editors

Andrea Magrini

Andrea Magrini is assistant professor in Fluid Machinery at the Department of Industrial Engineering of the University of Padova. He graduated in Aerospace Engineering and holds a PhD in Energetics. His research is focused on propulsive turbomachinery for marine and aviation applications, with a special focus on propulsion system integration. He is an expert on installation effects, propulsor/frame coupling, and the numerical modeling of component interactions. His work involves machine design, computational fluid dynamics modeling, innovative propulsive configurations, studies of energy efficiency, and advanced optimization. He has participated in several European research projects related to propulsion system development and optimization. He is the author of numerous peer-reviewed papers in international scientific journals and a member of ASME and AIAA.

Nastia Degiuli

Nastia Degiuli graduated in Naval Architecture from the Faculty of Mechanical Engineering and Naval Architecture, University of Zagreb, where she also obtained her MSc and PhD degrees. She is currently a Full Professor with tenure and Head of the Chair of Ship Hydrodynamics at the same institution. Her research focuses on numerical and experimental ship and offshore hydrodynamics, biofouling, and marine renewable energy. She has participated in numerous international and national collaborative research projects. Nastia Degiuli has authored a substantial body of work, including papers in highly ranked international journals, conference proceedings, and book chapters. She currently serves as Editor-in-Chief of *Brodogradnja: An International Journal of Naval Architecture and Ocean Engineering for Research and Development*, which is ranked in the first quartile (Q1) of the Web of Science (WoS). She is also a member of the editorial boards of several international journals, including *Ocean Engineering* and *Polish Maritime Research*. She is an active member of several scientific and professional organizations, including the Society of Naval Architects and Marine Engineers (SNAME), the International Towing Tank Conference (ITTC), and the Croatian Academy of Engineering.

Filippo Avanzi

Filippo Avanzi obtained his Master's degree in Aerospace Engineering from the University of Padova. He then earned his PhD in Energetics at the Department of Industrial Engineering of the same institution. He is currently an Assistant Professor in Fluid Machinery at the University of Padova, where his research focuses on marine propulsion fluid dynamics, with a particular emphasis on innovative configurations, machine/system coupling, and design optimization. He has been involved in several national projects related to waterjet propulsion and has published extensively on the topic in international peer-reviewed journals and conference proceedings. He has also served as Guest Editor for three MDPI journals and is a member of the American Society of Mechanical Engineers (ASME).

Preface

This reprint collects the contributions published in the Special Issue "Novelties in Marine Propulsion", addressing the growing demand for efficient and sustainable solutions in maritime transport. Motivated by increasingly stringent environmental regulations and technological challenges, this collection gathers recent advances in the design, modeling, and experimental investigation of propulsive systems. The works presented provide a comprehensive overview of current research directions, offering valuable insights for researchers and practitioners engaged in the development and optimization of next-generation marine propulsion technologies.

Andrea Magrini, Nastia Degiuli, and Filippo Avanzi
Guest Editors

Journal of
Marine Science
and Engineering

MDPI

Editorial

Novelties in Marine Propulsion

Andrea Magrini [1,*], Nastia Degiuli [2] and Filippo Avanzi [1]

[1] Department of Industrial Engineering, Università degli Studi di Padova, Office 00140 05 010-5th Floor, Via Venezia 1, 35131 Padova, Italy; filippo.avanzi@unipd.it
[2] Faculty of Mechanical Engineering and Naval Architecture, University of Zagreb, 10000 Zagreb, Croatia; nastia.degiuli@fsb.unizg.hr
* Correspondence: andrea.magrini@unipd.it

1. Introduction

The maritime industry is undergoing one of its most profound transitions since the shift from sail to steam, driven by increasingly stringent international regulations such as the International Convention for the Prevention of Pollution from Ships (MARPOL). Maritime transport supports over 80% of global trade and is therefore unavoidable in the current economic system [1]; at the same time, its environmental footprint is substantial and growing, with propulsion systems alone contributing up to 3% of global greenhouse gas (GHG) emissions according to recent International Maritime Organization (IMO) assessments [2]. Meeting the increasing demand for commercial shipping thus requires a fundamental redefinition of maritime transport in line with regulatory targets, including the IMO objective of halving GHG emissions by 2050 relative to 2008 levels. In this context, research has largely focused on energy supply, identifying fossil fuels—particularly heavy fuel oil—as the primary source of environmental impact. Consequently, an extensive body of literature has explored alternative energy carriers aimed at reducing GHG emissions, sometimes prioritizing decarbonization over combustion efficiency [3]. Among the most widely investigated solutions are battery-powered propulsion systems and biofuels [4–6], while nuclear energy is also emerging as a potential long-term option [7,8]. At the same time, the tailored design [9] and optimization [10] of propulsive units remain fundamental to fully exploit low-emission energy carriers and to further enhance system performance toward more sustainable maritime transport.

Driven by tightening regulatory constraints on maritime navigation, with the long-term objective of achieving net-zero emissions, the quest for "Novelties in Marine Propulsion" has evolved from a predominantly academic endeavor into a critical industrial necessity. To foster progress in sustainable marine propulsion, this Special Issue brings together eight pioneering studies that address these challenges through advanced Computational Fluid Dynamics (CFD), innovative design exploration, and the investigation of disruptive physical phenomena.

2. Published Articles

In the past decade, the reliance on traditional towing tank testing has been augmented—and in some design stages, replaced—by high-fidelity numerical simulations. The papers in this issue highlight the maturity of Unsteady Reynolds-Averaged Navier–Stokes (URANS) and Large Eddy Simulation (LES) techniques. As ship owners face stricter Energy Efficiency Existing Ship Index (EEXI) and Carbon Intensity Indicator (CII) regulations, the ability to predict hydrodynamic performance with sub-1% error margins (as seen in the works of Zhang et al. and Feng et al.) is feasible. These tools allow for the rapid "virtual prototyping" of

 https://doi.org/10.3390/jmse14080708

complex systems, such as the ducted waterjets and multi-propeller configurations discussed herein, which would be prohibitively expensive to iterate physically.

A recurring theme in this Special Issue is the shift from optimizing individual components to modelling the integrated propulsion system. Modern vessels are no longer characterized by a simple screw propeller; they are complex assemblies of ducts, stators, and interaction zones. The work by Feng et al. on body force modeling for waterjets emphasizes that as vessels grow in complexity, interference between multiple units becomes a dominant design factor. Wang et al. provide a systematic mapping of these interactions in a two-propeller configuration, offering concrete indications for spacing and arrangement to maximize collective efficiency.

Beyond the refinement of existing technologies, this issue explores groundbreaking novelties that challenge conventional naval architecture. The exploration of the Coanda effect by Lee et al. represents a step toward circulation-control propulsion, demonstrating that bio-inspired or fluid-dynamic phenomena can yield power reductions of nearly 8%. Moreover, the work by Avanzi et al. on submerged nacelles for the Outboard Dynamic-inlet Waterjet (ODW) introduces a new aero-derivative propulsive concept with the aim of extending the operational efficiency of unducted screw propellers and surpassing the installation penalties of traditional flush-type waterjets.

Finally, the transition to green fuels necessitates a reimagining of the vessel's internal architecture. The integration of Hydrogen Fuel Cells (PEMFCs), as analyzed by Chen et al., serves as a reminder that the propulsor of the future is only as efficient as the thermal and energy systems that support it. This comprehensive view—combining hydrodynamics, thermodynamics, and innovative physics—is the hallmark of the research presented in this Special Issue.

Summary of Published Papers

The study by Zhang et al. introduces a novel, simplified computational framework for assessing the hydrodynamic performance of ship propellers subjected to wave-induced heave and pitch motions. Addressing the critical limitation of traditional calm-water tests, the authors propose a hybrid methodology that integrates Potential Flow Theory (via Response Amplitude Operators, or RAOs) with Computational Fluid Dynamics (CFD) to model unsteady propeller forces. By prescribing heave and pitch motions derived from the potential flow method onto the propeller in a CFD environment, the research bypasses the high computational cost of full-ship–propeller coupling while maintaining engineering precision. Numerical simulations conducted on the KCS container ship and KP505 propeller reveal that vertical stern motions lead to significant performance degradation, specifically a thrust reduction and torque increase. Crucially, the results demonstrate an absolute efficiency decrease of 3 to 6 percentage points, representing an approximately 10% relative loss in propulsive performance under real sea conditions. Furthermore, the study identifies a significant nonlinear coupling between heave and pitch effects, indicating that their combined hydrodynamic impact is not merely a linear superposition of individual motions. Validated against high-fidelity CFD data, the proposed method achieves a four-fold reduction in computational time, making it a highly effective tool for systematic propulsion analysis in the early ship design phase and for the development of operational "Digital Twins" focused on energy efficiency.

The contribution by Park et al. provides a rigorous comparative analysis of numerical simulations for the open-water performance of cavitating propellers, a critical aspect in the design of high-speed General Purpose Planning Hulls (GPPHs). Utilizing two distinct computational codes—CFDShip-Iowa V5.5, an overset-based solver, and STAR-CCM+, a commercial general-purpose CFD tool—the authors evaluate the thrust and torque characteristics across a range of advance ratios (J) and cavitation numbers (σ). A pivotal finding of the study is the identification of significant performance degradation under low cavitation numbers (σ = 0.274), where averaged thrust, torque, and efficiency reductions reached approximately 63%, 50%, and 30%, respectively. The research highlights a technical discrepancy between the two solvers at high advance ratios: while CFDShip-Iowa captures complex blade-to-blade cavitation interactions and shedding, STAR-CCM+ (employing periodic boundary conditions for a single blade) tends to underpredict these violent patterns. This novelty in methodology underscores the necessity of full-propeller modeling over simplified periodic domains when simulating ultra-high-speed regimes where cavitation vapor may wrap around the entire assembly. By detailing the differences in vapor-phase iso-surfaces and pressure distributions, the work serves as a high-fidelity benchmark for naval architects aiming to refine the propulsive efficiency and noise signatures of high-speed craft. It ultimately emphasizes that for extreme operational conditions, the choice of mathematical model and boundary treatment is as decisive as the grid resolution in achieving predictive accuracy.

The study by Fraguela et al. presents a novel, data-driven framework for optimizing the operational efficiency of tugboats equipped with Voith–Schneider Propellers (VSPs). Unlike conventional screw propellers, VSP systems allow for independent control of both rotational speed (RPM) and blade pitch, providing 360-degree thrust vectoring but significantly increasing the complexity of energy management. To address the scarcity of full-scale empirical data in this field, the authors conducted extensive sea trials in the estuary of Ferrol, utilizing synchronized measurements of vessel speed, fuel consumption, and propulsion settings. The research introduces a methodology for generating interpolated fuel-consumption and efficiency maps using cubic interpolation and visual analytics, which identifies optimal operating "islands" segmented by speed ranges. A key contribution is the development of a relative efficiency indicator that allows masters to select the most energy-efficient combination of pitch and RPM for a required transit speed, bypassing the need for complex internal engine monitoring. By providing practical guidelines that could lead to significant fuel savings in port operations, this work serves as a foundational step toward the creation of Digital Twin architectures for harbor craft. It highlights that the novelty in modern propulsion research lies not only in hydrodynamic design but in the intelligent, data-led exploitation of existing high-maneuverability systems to meet the industry's decarbonization targets.

The research presented by Feng et al. addresses the computational bottleneck of simulating submerged waterjet propulsion systems by proposing an improved body force model. While high-fidelity simulations using the sliding mesh method provide an accurate representation of rotor–stator interactions, they are computationally intensive for full-scale ship applications. To bridge this gap, the authors developed a modified body force method that replaces the physical rotor and stator geometries with virtual cylindrical disks where forces are uniformly distributed. A key progress of this study is the introduction of a velocity correction procedure: the authors used the average flow velocity at the duct inlet to refine the open-water characteristic curves of thrust coefficient versus advance ratio, ensuring that the body force inputs reflect the actual internal ducting flow rather than an idealized external velocity. Validated against Experimental Fluid Dynamics (EFD) and sliding mesh results, the improved model demonstrated remarkable precision, with thrust coefficient errors as low as −0.61%. Furthermore, the study investigated the sensitivity of the inflow velocity plane position, concluding that any plane between the duct inlet and the propeller disk can be used for correction without compromising flow-field accuracy. This simplified numerical approach offers a robust balance between accuracy and efficiency, providing naval architects with a high-speed tool for analyzing the self-propulsion performance and hull–propulsion interaction of waterjet-equipped vessels.

The study by Avanzi et al. presents a comprehensive investigation into the design principles for optimal nacelle shaping in Outboard Dynamic-inlet Waterjets (ODWs). Recognizing the critical role of the intake geometry in delivering clean flow to propulsive pumps, the authors performed a multi-point design exploration using two-phase RANS simulations. A relevant feature of this work is the simultaneous assessment of performance at both cruise and take-off conditions, the latter specifically accounting for the potential onset of internal lip cavitation. By utilizing a B-spline parameterization to manipulate the nacelle's wall profiles, the research identifies key correlations between geometric features—such as the highlight-to-throat ratio—and propulsive metrics extracted via Thrust–Drag Bookkeeping (TDB). The findings emphasize that the Inlet Velocity Ratio (IVR) serves as a deterministic criterion for identifying operating regimes, showing how pre-diffusion or convergence in the capture streamtube directly affects the cavitation margin and internal efficiency. This multi-fidelity approach establishes ground rules for the geometric specification of submerged nacelles to ensure that next-generation propulsive pumps can operate efficiently across a wide range of vessel speeds. It bridges the gap between aero-engine research techniques and marine engineering, offering a sophisticated toolset for the optimization of outboard propulsion units.

The study by Wang et al. investigates the complex hydrodynamic interactions within a two-propeller configuration, offering design guidelines for multi-propeller propulsion systems. Utilizing an in-house viscous CFD code, the authors performed URANS simulations with a structured overset grid approach to analyze 45 different spatial arrangements of two KP505 propellers. The originality of this research lies in the systematic mapping of interference effects across a wide range of transverse and longitudinal spacings between the propellers. The results reveal that while the front propeller remains largely unaffected when spacing exceeds one diameter, the rear propeller suffers significant performance loss due to the high-velocity wake induced by the front unit. This interaction reduces its effective angle of attack, leading to substantial decreases in thrust and torque. Interestingly, the study finds that longitudinal spacing has a negligible effect on the rear propeller due to the minimal decay of wake velocity, whereas transverse spacing is the dominant factor. The authors conclude that to achieve optimal overall propulsive efficiency, surpassing that of a single propeller, the transverse spacing must be maintained at no less than one diameter. This comprehensive dataset provides a valuable benchmark for the arrangement of twin-screw vessels and the development of energy-efficient multi-propulsor architectures in modern naval architecture.

Chen et al. addressed a critical challenge in the adoption of hydrogen-powered marine vessels: the thermal–hydraulic management of Proton Exchange Membrane Fuel Cells (PEMFCs). Efficient cooling is vital for maintaining fuel cell longevity and performance, yet conventional cooling-plate designs often suffer from high pressure drops or uneven temperature distributions. The authors investigated six different cooling channel configurations (Types A to F) using COMSOL Multiphysics simulations to optimize heat dissipation. The research introduces secondary flow channels designed to enhance fluid mixing and reduce thermal gradients. Through a rigorous orthogonal experimental design, the researchers identified that channel width is the dominant factor influencing both temperature uniformity and pressure drop. The optimized Type B configuration, featuring these secondary channels, demonstrated a significant improvement in the cooling efficiency coefficient, balancing the trade-off between parasitic pumping power and heat removal capacity. By providing a detailed sensitivity analysis of channel angles and spacing, this work offers a robust engineering blueprint for the thermal design of high-power fuel cell stacks. It

https://doi.org/10.3390/jmse14080708

underscores that the transition to green propulsion requires not only new propulsors but also the optimization of the auxiliary systems that sustain zero-emission energy converters in the demanding marine environment.

The final contribution by Lee et al. introduces a transformative approach to marine propulsion by integrating the Coanda effect into propeller design to enhance self-propulsion efficiency. Using a 6.5K DWT tanker as the baseline model, the authors conducted a series of numerical simulations using URANS equations and the SST k–ω turbulence model to compare three configurations: an original propeller, a standard propeller, and a specialized Coanda propeller. In the latter, the Coanda effect was used to generate additional lift and thrust even at significantly lower rotational speeds. The study's results are compelling, demonstrating that the Coanda-based system achieved a reduction in delivered power of approximately 7.8% compared to traditional propulsion setups. This improvement in the self-propulsion factor not only suggests a path toward more fuel-efficient vessel operations but also offers a viable alternative for noise and vibration reduction in sensitive maritime environments. By providing a rigorous numerical baseline for this technology in a ship-scale context, the authors pave the way for a new generation of high-efficiency, bio-inspired, or circulation-control propulsors that challenge the conventional limits of screw propeller theory.

Conflicts of Interest: The authors declare no conflicts of interest.

List of Contributions:

1. Lee, J.-H.; Paik, K.-J.; Oh, D.-H. A Numerical Study for the Self-Propulsion Performance of a Propulsion System Using the Coanda Effect. *J. Mar. Sci. Eng.* **2025**, *13*, 437. https://doi.org/10.3390/jmse13030437.
2. Chen, W.; Yin, Y.; Liu, Y. Numerical Analysis and Optimization of Secondary Flow Channels in a PEMFC Cooling Plate. *J. Mar. Sci. Eng.* **2025**, *13*, 764. https://doi.org/10.3390/jmse13040764.
3. Wang, X.; Ma, Y.; Feng, D.; Yao, C.; Cai, Z. Numerical Study of the Hydrodynamic Performance of a Two-Propeller Configuration. *J. Mar. Sci. Eng.* **2025**, *13*, 992. https://doi.org/10.3390/jmse13050992.
4. Avanzi, F.; Magrini, A.; De Vanna, F. Two-Phase Multi-Point Design Exploration of Submerged Nacelles for Marine Propulsive Pump Installation. *J. Mar. Sci. Eng.* **2025**, *13*, 1110. https://doi.org/10.3390/jmse13061110.
5. Feng, D.; Ma, Y.; Cai, Z.; Yang, P.; Zou, Y. A Modified Body Force Model for a Submerged Waterjet. *J. Mar. Sci. Eng.* **2025**, *13*, 1314. https://doi.org/10.3390/jmse13071314.
6. Fraguela, F.; Mendizábal, F.; Pérez-Canosa, J.M.; Orosa, J.A. Data-Driven Optimization of Voith-Schneider Tug Operations: Towards a Digital Twin Framework for Port Energy Management. *J. Mar. Sci. Eng.* **2025**, *13*, 1405. https://doi.org/10.3390/jmse13081405.
7. Park, S.; Wang, Z.; Stern, F.; Gunderson, A.; Scherer, J. Comparison of Numerical Simulations of Propeller Open-Water Performance with Cavitation for High-Speed Planing Hulls. *J. Mar. Sci. Eng.* **2025**, *13*, 1804. https://doi.org/10.3390/jmse13091804.
8. Zhang, W.; Xu, G.; Wang, S.; Wu, R. A Simplified Prediction Method for the Hydrodynamic Performance of Ship Propellers Under Heave and Pitch Motions. *J. Mar. Sci. Eng.* **2025**, *13*, 2041. https://doi.org/10.3390/jmse13112041.

References

1. Lee, S.-S. Evaluating the safety of the redesigned shafting system following the propeller redesign aimed at enhancing CII and EEXI. *Results Eng.* **2024**, *23*, 102352. [CrossRef]
2. International Maritime Organization. Fourth IMO GHG Study 2020. 2021; pp. 1–2. Available online: www.imo.org (accessed on 7 April 2026).
3. Delgarm, N.; Varnousfaaderani, M.R.; Farrokhfal, H.; Ardeshiri, S. Thermo-mathematical modeling and multi-objective optimization of marine propulsion using artificial neural network and a novel physics-based metaheuristic optimization algorithm: Thermo-economic-environmental analysis. *Nucl. Eng. Technol.* **2025**, *57*, 103601. [CrossRef]

https://doi.org/10.3390/jmse14080708

4. Manikandan, S.; Vickram, S.; Devarajan, Y. Cutting-edge technologies: Biofuel innovations in marine propulsion systems to lower black carbon emissions. *Results Eng.* **2025**, *25*, 104095. [CrossRef]
5. Ünlübayir, C.; Youssfi, H.; Khan, R.A.; Ventura, S.S.; Fortunati, D.; Rinner, J.; Börner, M.F.; Quade, K.L.; Ringbeck, F.; Sauer, D.U. Comparative analysis and test bench validation of energy management methods for a hybrid marine propulsion system powered by batteries and solid oxide fuel cells. *Appl. Energy* **2024**, *376*, 124183. [CrossRef]
6. Chountalas, T.D.; Founti, M.; Hountalas, D.T. Review of Biofuel Effect on Emissions of Various Types of Marine Propulsion and Auxiliary Engines. *Energies* **2023**, *16*, 4647. [CrossRef]
7. Xu, Z.; Liu, J.; Chen, Y.; Li, A. Numerical Simulation of Environmental Characteristics of Containment in Severe Accident of Marine Nuclear Power Plant. *Energies* **2023**, *16*, 7821. [CrossRef]
8. Ma, X.; Alfani, D.; Silva, P.; Sun, B. Preliminary design and optimization of sCO_2 nuclear-powered merchant vessel propulsion systems for the Northern Sea Route. *Energy Convers. Manag.* **2025**, *345*, 120287. [CrossRef]
9. Sulovsky, I.; Mewes, S.; El Moctar, O.; Prpić-Oršić, J. CFD study on a full-scale ship performance in a representative sea state. *Ocean. Eng.* **2025**, *333*, 121519. [CrossRef]
10. Bacalja Bašić, B.; Krčum, M.; Jurić, Z. Propeller Optimization in Marine Power Systems: Exploring Its Contribution and Correlation with Renewable Energy Solutions. *J. Mar. Sci. Eng.* **2024**, *12*, 843. [CrossRef]

https://doi.org/10.3390/jmse14080708

Journal of
Marine Science and Engineering

MDPI

Article

A Simplified Prediction Method for the Hydrodynamic Performance of Ship Propellers Under Heave and Pitch Motions

Wei Zhang [1,*], Guoliang Xu [1], Siyu Wang [1] and Rui Wu [2]

[1] School of Ocean Engineering, Harbin Institute of Technology (Weihai), Weihai 264209, China
[2] Shanghai Ship and Shipping Research Institute Co., Ltd., Shanghai 200135, China
* Correspondence: zhang.wei@hit.edu.cn

Abstract: This paper presents a simplified computational method for evaluating the hydrodynamic performance of a ship propeller, taking into account the effects of wave-induced heave and pitch motions. The method combines Computational Fluid Dynamics (CFD) for propeller hydrodynamic force calculations with forced heave and pitch motions computed via the potential flow theory-based method. Numerical simulations at three representative wavelengths demonstrate that heave and pitch motions reduce propeller thrust, increase torque, and decrease the propeller's efficiency significantly. In addition, the results reveal a nonlinear coupling between heave and pitch effects, where their hydrodynamic contributions are not simply additive. Compared to high-fidelity ship–propeller coupling CFD simulations, the proposed method reduces computational costs by approximately four times while maintaining practical accuracy for wave–propeller interaction analysis.

Keywords: propeller hydrodynamics; heave and pitch motions; ship propulsion in waves; KCS; KP505 propeller

1. Introduction

To address global environmental concerns, the International Maritime Organization (IMO) has implemented stringent regulations including the Energy Efficiency Design Index (EEDI) for new-built ships [1] and the Carbon Intensity Indicator (CII) for existing fleets [2] Moreover, IMO has set a target to reduce greenhouse gas emissions from international shipping by at least 50% by 2050 compared to 2008 levels, with an aim to phase out emissions entirely within this century [3].

In this context, energy conservation and emission reduction have become the most important trends in the development of the shipping industry. Accordingly, there has been a growing shift in the field of marine engineering from traditional studies of ship performance in calm water to research on ship performance in waves, aiming to enhance navigation efficiency under real sea conditions.

Screw propellers (referred to henceforth, in keeping with common practice, as propellers) are the most common propulsion devices for ships. While traditional evaluation methods of propeller hydrodynamics (open-water and behind-ship tests) remain valuable, their inability to account for wave effects represents a critical limitation in real-world performance prediction. To enhance a ship's performance in actual sea conditions, it is deemed important to study the hydrodynamic performance of propellers in waves. Tavakoli et al. [4] and Saettone et al. [5], through their study on mathematical models of ship propulsion systems in waves, confirm that the variation in propeller performance under wave conditions is a crucial and complex component of the entire propulsion system.

Recent studies on propeller performance in waves primarily adopt two approaches: open-water analyses with wave effects consideration and coupled hull-propeller analyses under wave conditions. Irannezhad et al. [6] conducted experimental and numerical open water tests of KP458 propellers in calm water and in regular head waves, demonstrating the remarkable influences of waves on the propeller advance coefficient. Recently, Acanfora et al. [7] conducted experimental studies on the open water characteristics of a ship propeller in waves and showed that both thrust and torque are reduced compared to the calm water case as wave amplitude increases. Cosgun et al. [8] proposed a method to predict the motion performance of a submarine under various surge conditions by utilizing four-quadrant propeller performance and resistance test data. Their work also enhanced the prediction scope for the propeller's open-water performance. Liang et al. [9] analyzed the open-water propulsion performance of a propeller and the scale effect on hull resistance. They demonstrated that although the extrapolation method from model to full scale lacks high accuracy, its relative error remains acceptable for engineering requirements. Yurtseven et al. [10] investigated the variation pattern of blade torque during the pitch change action of a controllable pitch propeller. Their results show that the torque required to rotate each blade increases with the pitch angle, peaks at a critical point, and then decreases. Grlj et al. [11] conducted an in-depth investigation into the scale effects on open-water propeller performance. Through a systematic and scientific study of propeller open-water tests with and without the Gamma-Re-θ transition model, they found that the influence of scale effects on open-water characteristics was significantly mitigated when the transition model was applied.

Analyzing the performance of a propeller in ship behind conditions is more complex and intriguing than in open water tests. Particularly, when the ship is sailing in waves, the propeller experiences six degrees of freedom wave-induced motions along with the ship hull. These significant wave-induced motions alter the wake of the ship hull, further complicating the flow field around the propeller. Through model testing, Ueno et al. [12,13] reveal that heave and pitch motions alter propeller inflow, leading to measurable changes in hydrodynamic performance.

Thanks to the rapid advancements in Computational Fluid Dynamics (CFD), numerous studies have also been conducted in this area. For instance, Cai et al. [14] tested three different methods to analyze propeller–hull interactions in head waves. Using the KCS ship as a research model, Gamal et al. [15] analyzed the variation in propeller propulsive power behind a ship in different wave conditions and demonstrated that an increase in wave steepness leads to large fluctuations in thrust, resulting in an additional reduction in propeller efficiency. Woeste et al. [16] found that added power is not simply a linear function of added resistance or the squared incoming wave amplitude; the complex interactions between the propeller and hull in the waves result in significant power consumption. Sadat-Hosseini et al. [17] simulated the self-propelled KCS ship in regular head waves with a dimensionless wavelength of $\lambda/L = 1.37$ and confirmed a substantial increase in the mean inflow velocity to the propeller due to the fluctuating wake field. Ortolani et al. [18] studied the loads on individual blades of the propeller of a twin-screw ship model navigating in regular waves, demonstrating that single blade loads are significantly affected by waves, even in relatively moderate sea states (*SS3* and *SS4*).

Indeed, the above-mentioned CFD studies provide deep understanding of the complex physical phenomena of hull-propeller-wave interactions. However, these computations require large computational resources and long CPU times and, therefore, it is hard to perform systemic computations that reveal the variation law of the hydrodynamic performance of the propeller, such as the thrust and torque against the incident wave conditions. To deal with this problem, Wang et al. [19,20] proposed a simplified method by simulating a propeller in open water conditions near the free surface, with a forced sinusoidal heave motion applied to the propeller shaft. Their results proved that the propeller load and the wake field are closely related to the variation of heave motion period.

We computed the hydrodynamic performance of the KP505 propeller considering the heave and pitch motions of the KCS hull. A CFD-based numerical method was developed for the primary propeller analysis, and a potential flow-based method was used to calculate the response amplitude operators (RAOs) of the heave and pitch motions of the ship advancing in waves. Referring to Wang et al. [19] we exerted sinusoidal heave and pitch motions, which are generated according to the computed amplitudes and phase lags of RAOs, to the propeller. Based on the developed model, three different incident wave lengths were tested and the propeller thrust, torque and efficiency under different wave conditions were collected and analyzed. The calculated results are expected to provide theoretical reference for the estimation of propeller propulsion performance in waves.

2. Computational Method

2.1. Numerical Simulation Method of CFD

Considering an incompressible viscous fluid, its flow adheres to the principles of mass and momentum conservation. Accordingly, the fluid's continuity equation and momentum equations (Navier-Stokes equations) read as follows:

$$\frac{\partial\rho}{\partial t} + \nabla\cdot(\rho\mathbf{v}) = 0 \tag{1}$$

$$\frac{\partial(\rho\mathbf{v})}{\partial(t)} + \nabla\cdot(\rho\mathbf{v}\otimes\mathbf{v}) = -\nabla\cdot(p\mathbf{I}) + \nabla\cdot\mathbf{T} + \mathbf{f}_b \tag{2}$$

where ρ is the fluid density, $\mathbf{v}$ is the fluid velocity vector, t is time, p is pressure, $\mathbf{I}$ is the unit stress tensor, $\mathbf{T}$ is viscous stress tensor, and $\mathbf{f}_b$ is the body force acting on the fluid.

For most practical engineering applications, fluid flow is typically turbulent, marked by irregular and highly rotational motion. These turbulent structures are small in scale and high in frequency. Employing Direct Numerical Simulation (DNS) to resolve them is prohibitively expensive and time-consuming, making it infeasible for routine use. Conversely, the Reynolds-Averaged Navier–Stokes (RANS) method has found extensive application in ship hydrodynamics and related fields because of its relatively high computational efficiency. The RANS method decomposes the solution variables in the instantaneous Navier–Stokes equations into mean and fluctuating components. Substituting these decomposed variables into Equations (1) and (2) yields the following time-averaged continuity and momentum equations:

$$\frac{\partial\rho}{\partial t} + \nabla\cdot(\rho\overline{\mathbf{v}}) = 0 \tag{3}$$

$$\frac{\partial(\rho\overline{\mathbf{v}})}{\partial(t)} + \nabla\cdot(\rho\overline{\mathbf{v}}\otimes\overline{\mathbf{v}}) = -\nabla\cdot(\overline{p}\mathbf{I}) + \nabla\cdot(\overline{\mathbf{T}} + \mathbf{T}_{RANS}) + \mathbf{f}_b \tag{4}$$

where $\overline{\mathbf{v}}$ is the mean velocity, $\overline{p}$ is the mean pressure, $\mathbf{T}_{RANS}$ is the additional stress tensor, its expression is:

$$\mathbf{T}_{RANS} = -\rho\mathbf{R} + \frac{2}{3}\rho k\mathbf{I} \tag{5}$$

where k is the turbulent kinetic energy, $\mathbf{R}$ is the Reynolds stress tensor, which is defined as:

$$\mathbf{R} = \begin{bmatrix} \overline{u'u'} & \overline{u'v'} & \overline{u'w'} \\ \overline{u'v'} & \overline{v'v'} & \overline{v'w'} \\ \overline{u'w'} & \overline{v'w'} & \overline{w'w'} \end{bmatrix} \tag{6}$$

The addition of the extra stress tensor $\mathbf{T}_{RANS}$ makes the original governing equations unclosed. Thus, a turbulence model is required to model this stress and achieve closure. In this work, the Reynolds stress is modeled using the eddy viscosity concept, which relates it to the mean flow field via the Boussinesq approximation as follows:

$$\mathbf{T}_{RANS} = 2\mu_t \mathbf{S} - \frac{2}{3}(\mu_t \nabla \cdot \overline{\mathbf{v}})\mathbf{I} \tag{7}$$

where $\mathbf{S}$ is the mean strain-rate tensor defined as:

$$\mathbf{S} = \frac{1}{2}(\nabla\overline{\mathbf{v}} + \nabla\overline{\mathbf{v}}^T) \tag{8}$$

To model the turbulent eddy viscosity (μ_t), the Shear Stress Transport (SST) k-ω model is adopted. This approach merges the k-ω and k-ε models by substituting ω for ε in the k-ε model via the SST equations, effectively combining the far-field behavior of k-ε with the near-wall accuracy of k-ω. SST k-ω model transport equations:

$$\frac{\partial(\rho\omega)}{\partial t} + \nabla \cdot (\rho\omega\overline{\mathbf{v}}) = G_\omega + S_\omega + D_\omega + \nabla \cdot [(\mu + \sigma_\omega \mu_t)\nabla\omega] - \rho\beta f_\beta(\omega^2 - \omega_0^2) \tag{9}$$

$$\frac{\partial(\rho k)}{\partial t} + \nabla \cdot (\rho k\overline{\mathbf{v}}) = G_k + S_k + \nabla \cdot [(\mu + \sigma_k \mu_t)\nabla k] - \rho\beta^* f_{\beta^*}(\omega k - \omega_0 k_0) \tag{10}$$

where ω is specific dissipation rate, μ is dynamic viscosity, σ_k and σ_ω are model coefficients, G_k, D_k and D_ω are production terms, f_{β^*} is free shear correction factor, f_β is vortex stretching modification factor, S_ω and S_k are source terms.

2.2. General Methodology

Let us consider a ship advancing at a constant forward speed, Vs, in regular head waves. The propeller, operating in the behind-hull condition, generates thrust to overcome the total resistance, which is the sum of the calm-water resistance and the added wave resistance. Note that both the propeller thrust and the wave-added resistance are inherently variable rather than constant due to the influence of the waves. However, to avoid the complexities of the nonlinear interactions between thrust, ship speed, and wave-added resistance, we introduce a simplified computational method. Firstly, given the incident wave conditions and the fixed ship speed, we compute heave and pitch RAOs based on potential flow theory. Secondly, we ignore the influence of wave-induced ship motions on propeller thrust and evaluate the 'ideal' rotation speed of propeller based on its open water characteristics in calm water. Thirdly, we enforce the heave of pitch motions on the open-water propeller and analyze the related variations of hydrodynamic performance of the propeller.

In practice, the ANSYS (2023 R2) AQWA software is applied to compute the motion RAOs. AQWA is a numerical analysis tool based on potential flow theory. It computes the global motion responses of ships to wave loads with high computational efficiency. The following sections introduce the methods for determining the 'ideal' propeller rotation speed and analyzing propeller performance.

2.3. Approach for Identifying the 'Ideal' Rotation Speed

The 'ideal' rotation speed of propeller is determined by solving the system of equations consisting of (11) to (14):

$$(1-t)T = R_c + R_w \tag{11}$$

$$T = K_T(J) \cdot \rho n^2 D^4 \tag{12}$$

$$J = \frac{V_A}{nD} \quad (13)$$

$$V_A = (1 - \omega_0)V_S \quad (14)$$

where t is the thrust deduction factor, Rc and Rw represent the calm water resistance and the mean value of the wave added resistance, respectively. T is the 'ideal' thrust force of the propeller, n is the 'ideal' rotation speed; D is the diameter of the propeller, ω_0 is wake coefficient and K_T is the open water thrust coefficient of the propeller in calm water. Equation (11) represents the equilibrium between propeller thrust and hull resistance, whereas Equations (12)–(14) are used to calculate the rotational speed of the propeller behind the ship under self-propulsion conditions.

Following conventional methods for predicting ship navigation motion in waves, such as Lee et al. [21] and Okuda et al. [22], we assume that the thrust deduction factor and the wake coefficient remain constant regardless of the incident wave conditions. These parameters are determined through traditional self-propulsion tests conducted in calm water. By utilizing the calm water resistance, the mean wave-added resistance and the open-water characteristics of the propeller, we can determine the 'ideal' rotational speed of the propeller, namely, n, by solving Equations (11)–(14). This 'ideal' rotational speed theoretically ensures the ship maintains a constant forward speed, Vs, while sailing in waves.

2.4. CFD-Based Simulation Method for Propeller Performance Analysis

The hydrodynamic characteristics of propeller are computed using the CFD platform STAR-CCM+ (2410 19.06.008). STAR-CCM+ is a widely used CFD tool in ship hydrodynamics. Its built-in meshing capabilities are robust. Furthermore, its user-defined field functions and superposed motion feature are ideally suited for applying complex prescribed motions to components. A computational domain, which comprises three zones, is employed. The innermost zone simulates propeller motion (rotation and heave/pitch). The intermediate zone defines the range of heave/pitch motion and employs mesh refinement to enhance flow-field prediction accuracy. The outermost zone represents the overall fluid domain. Figure 1 illustrates the dimensions of each zone. An overlapping grid interface is set up between the innermost zone and the intermediate zone.

In order to improve the computational efficiency, we consider only the open-water condition in our simplified method. Nevertheless, the influence of the ship hull is modelled via the wake coefficient. The advance speed of the propeller, V_A, is applied as a velocity-inlet boundary condition at the frontal boundary of the outermost zone. The setup methods for other boundary conditions are shown in Figure 1.

In order to model the propeller's working conditions during heave and pitch motions, prescribed sinusoidal heave and pitch movements were applied as follows:

$$y_h = A_h \sin(\omega t + \varphi_h) \quad (15)$$

$$y_p = \theta_p \sin(\omega t + \varphi_p) \quad (16)$$

where A_h and θ_p are the amplitude of heave and pitch motions, respectively; ω is the motion frequency, which equals to the wave encounter frequency; φ_h and φ_p denote phase lags of heave and pitch motions with respect to the incident wave profile, respectively. Figure 2 shows the schematic definition of heave and pitch motions. In practice, the motion amplitude and phase lags are determined according to the seakeeping results provided by AQWA.

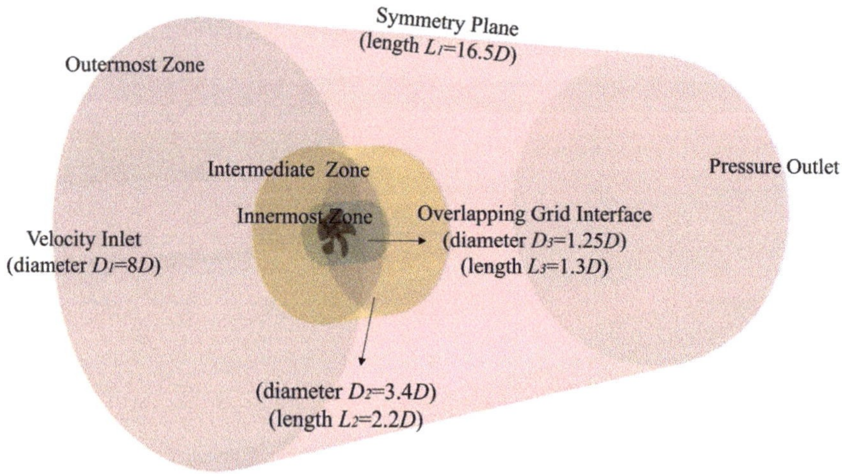

Figure 1. Schematic diagram of computational domain decomposition.

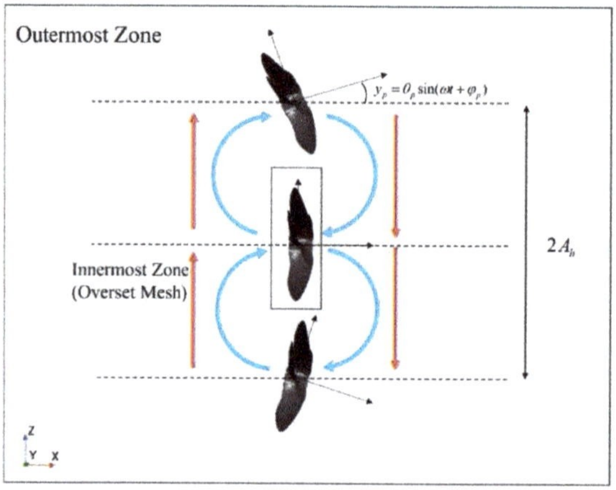

Figure 2. Schematic diagram of the heave and pitch motions of the propeller.

3. Numerical Setup

3.1. Test Cases

Numerical investigations are performed for the KCS container ship and its matched KP505 propeller. Designed by Korea Research Institute of Ships and Ocean Engineering (KRISO), the KCS container ship is a standard benchmark model widely used in ship hydrodynamics research. Extensive publicly available data exists for this model, including motion RAOs, wave-added resistance, and open-water propeller performance curves, providing comprehensive reference data for validation in this study. Figure 3 displays the KCS hull geometry, with principal particulars listed in Table 1. The KP505 propeller is the designated propeller for the KCS hull. The geometry of this propeller is illustrated in Figure 4, while its principal parameters are presented in Table 2.

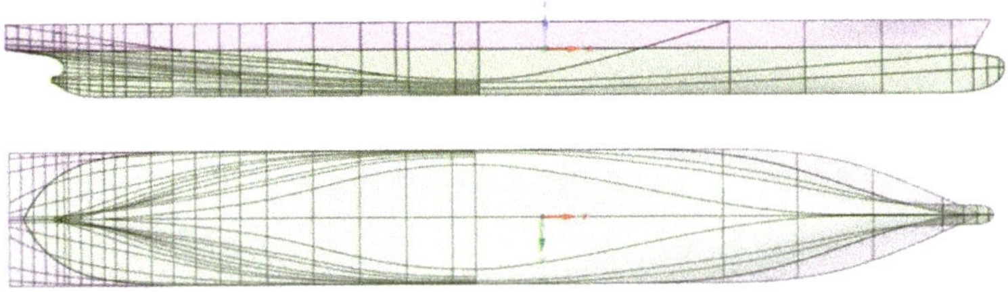

Figure 3. KCS ship model.

Table 1. KCS hull principal parameters.

Parameters	Value
Length L	230 m
Beam B	32.2 m
Displacement volume ∇	52,030 m^3
Draft T	10.8 m
Moment of Inertia K_{xx}/B	0.40
Moment of Inertia K_{yy}/L	0.25
Moment of Inertia K_{zz}/L	0.25

Figure 4. KP505 propeller.

Table 2. Propeller principal parameters.

Parameters	Value
Diameter D	0.250 m
Number of blades	5
P/D (0.7R)	0.997
Rotation direction	right-handed

3.2. Grid Generation Method

To conduct the hydrodynamic analysis of the propeller using CFD, a hybrid mesh generation approach was employed to discretize the flow domain. Specifically, the outermost zone and the intermediate zone were discretized using trimmed cell mesh and polyhedral mesh. The innermost zone, which is of high geometric complexity, was discretized using polyhedral mesh. Referring to Stark et al. [23], auxiliary volume control was implemented to achieve a smooth transition to the intermediate zone, with the transitional mesh detailed in Figure 5. The boundary layer was resolved with prismatic layers. This employed a low y^+ wall treatment strategy, targeting a value of $y^+ = 1$ for the first cell layer. Furthermore, geometric consistency was maintained between the outermost prism layer and the dimensions of the background mesh. The number of prism layers was set to 10, and the volume growth rate was set to 1.2. Figure 6 illustrates the detailed prism layer mesh.

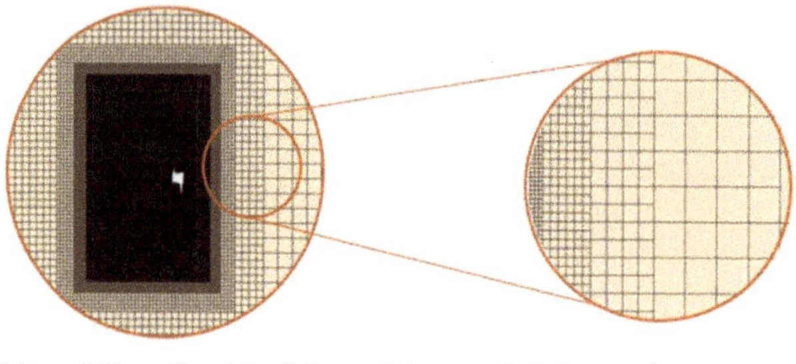

Figure 5. Transitional mesh between intermediate zone and outermost zone.

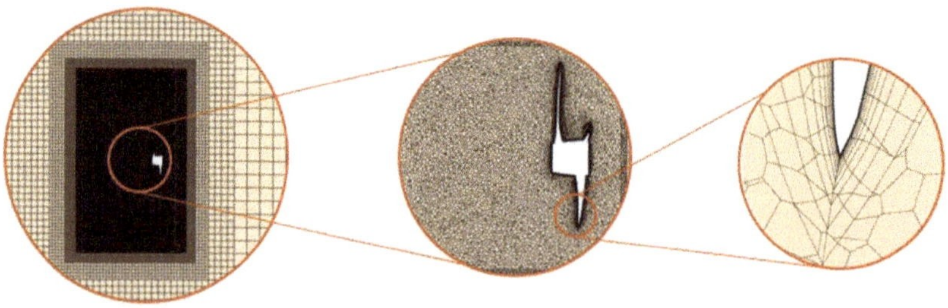

Figure 6. Prism layer mesh of propeller.

4. Results and Discussions

Numerical simulations were performed for a 1/31.6 scale model of the ship and propeller. A constant ship forward speed, corresponding to a Froude number $Fr = V_s/(gL)^{0.5} = 0.26$, was set as the target speed. The RAOs for heave and pitch motions at this speed were computed using AQWA. Based on the results, the heave and pitch motions under three representative wave conditions were selected to study their effects on the propeller's hydrodynamic performance.

This section first presents the RAOs results calculated using AQWA. Then, the open-water characteristics of the KP505 propeller in calm water, predicted by CFD computations, is presented and compared against the experimental measurements. Subsequently, the proposed simplified method is verified through comparison with coupled hull-propeller-wave CFD simulation results. Finally, the hydrodynamic performances of the KP505 propeller, under different heave and pitch motion conditions, are presented and discussed.

4.1. Numerical Results of Heave and Pitch RAOs of the KCS Ship in Head Waves

The seakeeping analysis of the KCS ship is performed using AQWA. The computed RAOs and phases of the heave and pitch motions are plotted in Figure 7. In Figure 7a,b, our numerical results (Present) are compared against the experimental measurements (EFD) and existing potential flow solutions (PF) of Sadat-Hosseini et al. [17]. As seen, the computed motion RAOs exhibit favorable agreement with the published experimental data and the existing numerical solutions, demonstrating the effectiveness of the potential flow methodology implemented in our study.

4.2. CFD Predictions of the Open-Water Characteristics of the KP505 Propeller in Calm Water

The hydrodynamic characteristics of the KP505 propeller are computed using STAR-CCM+. The propeller rotation is modelled via overset grid method. During each CFD simulation run, the prism layer thickness and number of layers were adjusted based on the propeller's advance speed to ensure computational accuracy.

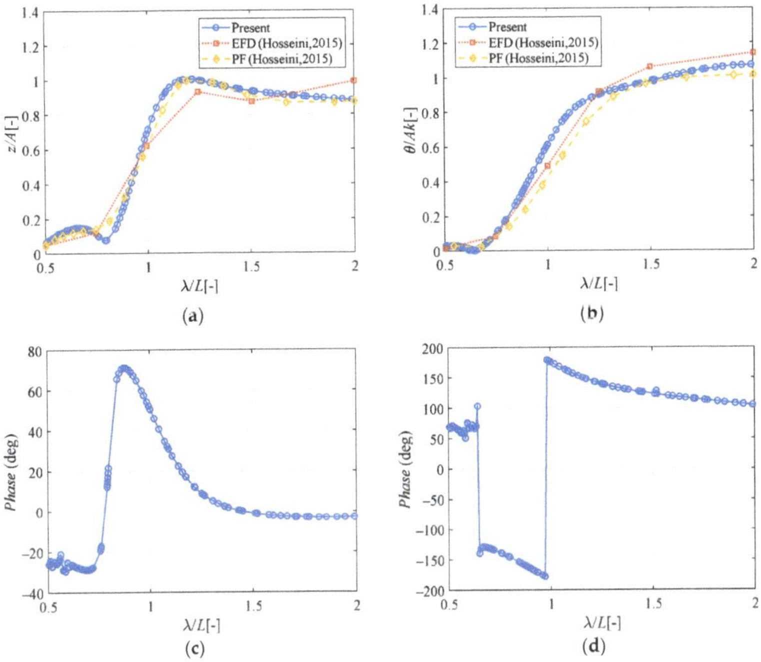

Figure 7. Heave and pitch RAOs and phases for the KCS ship advancing at Fr = 0.26 in head waves [17]. (**a**) Heave RAO; (**b**) Pitch RAO; (**c**) Phase of heave motion; (**d**) Phase of pitch motion.

Figure 8 plots the computed thrust coefficient (K_T), torque coefficient (K_Q), and propeller efficiency (η), for KP505 propeller in calm water. The tested advance ratio was in the region of $0.1 < J < 0.9$. Our numerical results (Present) are denoted by dashed line with hollow symbols while the comparative measurements (EFD), borrowed from Lee et al. [24], by solid lines with colored symbols.

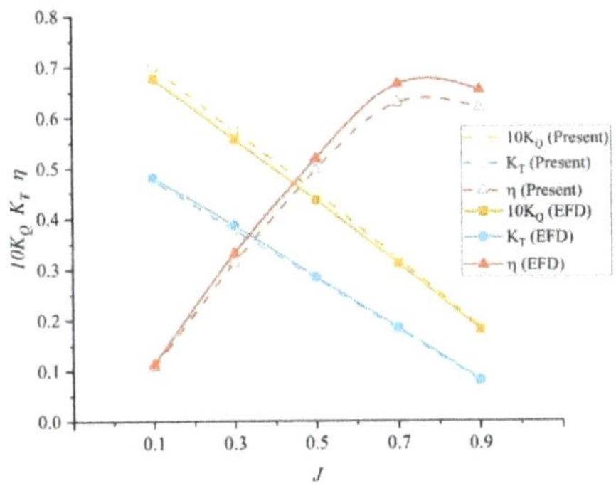

Figure 8. Open-water characteristics of the KP505 propeller in calm water.

It can be seen that our CFD results are in good agreement with the experimental results. The overall error is less than 5%. Near the design working condition (J = 0.5), the errors for the propeller thrust coefficient and torque coefficient are approximately 2%. The favorable agreements demonstrate the effectiveness of the established numerical approach for the propeller hydrodynamic analysis.

4.3. Hydrodynamic Performance of the Propeller with Heave and Pitch Motions

After validating the propeller performance, the focus shifts to added resistance under regular wave conditions. As seen in Figure 7, the peak value of the heave RAO occurs around a dimensionless wavelength (λ/L) is 1.15. Meanwhile, the pitch RAO increases significantly with wavelength when the dimensionless wavelength is below 1.15, while its growth rate decreases noticeably beyond this value. Hence, we selected three representative wavelengths, which centered around dimensionless wavelength is 1.15, as the typical computational cases. The default wave steepness (H/λ) is set to be 1/60.

Tables 3 and 4 summarize the detailed specifications of the selected computational cases, in which the motion RAOs are sampled from Figure 7 and the 'ideal' rotation speed for each case is calculated according to Equations (11)–(14). Note that a constant calm water resistance, Rc = 84 N, was considered for all tested cases. This value is computed via a CFD computation of the KCS ship model advancing in calm water at Fr = 0.26. The wave-added resistance coefficient σ_{aw} at different wave lengths are borrowed from the experimental measurements of Mikkelsen et al. [25] Moreover, a constant wake fraction, ω = 0.2071, and a constant thrust deduction coefficient, t = 0.1599, were adopted for all tested cases. These values were determined by Song et al. [26] based on the experimental self-propulsion tests in calm water.

Table 3. Propeller motions under different wavelength conditions.

Dimensionless Wavelength λ/L	The Amplitude of Heave Motion RAO	The Amplitude of Pitch Motion RAO	The Phase of Heave Motion (Rad)	The Phase of Pitch Motion (Rad)
1.00	0.714	0.613	0.8800	3.0818
1.15	0.996	0.844	0.3397	2.6240
1.30	0.984	0.922	0.0906	2.3532

Table 4. Hull and propeller parameters at different test cases (at 1/31.6 model scale).

Dimensionless Wavelength λ/L	Added Resistance Coefficient σ_{aw}	Total Resistance R (N)	Advance Coefficient J	'Ideal' Propeller Rotation Speed n (rps)
1.00	7.94	155.42	0.557	12.490
1.15	9.88	161.59	0.519	13.400
1.30	7.56	158.66	0.522	13.318

4.3.1. Convergence Test

Before conducting systemic computation of the propeller performance using the proposed method, we first carried out grid and temporal convergence test, following the methodology of Stern et al. [27] and Grlj et al. [28]. Selecting a typical case with dimensionless wavelength of 1.15, we computed the propeller thrust coefficient K_T and torque coefficient K_Q using three grid sets: coarse (S_{G1}, 5.23 million cells), medium (S_{G2}, 6.88 million cells), and fine (S_{G3}, 9.15 million cells). The grid refinement ratio Γ_G was approximately 1.31. Details about mesh density appear in Table 5.

Table 5. Details of mesh distribution.

Domain	Coarse	Medium	Fine
Outermost Zone	0.77 M	1.01 M	1.34 M
Intermediate Zone	4.28 M	5.50 M	7.32 M
Innermost Zone	0.27 M	0.37 M	0.49 M
Total	5.23 M	6.88 M	9.15 M

Similarly, temporal convergence was evaluated using three kinds of time steps, including 4.14×10^{-4} s (S_{T1}, 2° per time step), 2.07×10^{-4} s (S_{T2}, 1° per time step), and 1.03×10^{-4} s (S_{T3}, 0.5° per time step). This gave a time step refinement ratio $\Gamma_T = 2$. Based on the spatial and temporal discretization schemes, the expected order of accuracy (P_{Gest}) for grid convergence was 2. For time step convergence (P_{Test}), it was 1. The grid convergence study used the medium time step (S_{T2}). Conversely, the time step convergence study employed the medium grid (S_{G2}). Results for grid convergence and time step convergence are presented in Tables 6 and 7, respectively.

Table 6. Results of grid convergence validation ($\lambda/L = 1.15$).

	$\varepsilon_{23}\%S_{G3}$ (a)	$\varepsilon_{21}\%S_{G3}$ (b)	R_G (c)	Con. (d)	P_G (f)	C_G (g)	$U_G\%S_{G3}$ (h)	$U_I\%S_{G3}$ (i)
K_Q	0.3824	0.2689	0.7032	Mon. (e)	1.2345	0.5488	0.4088	0.0109
K_T	0.3522	0.2726	0.7741	Mon.	0.8977	0.3795	0.7490	0.0168

(a) $\varepsilon_{23}\%S_{G3}$ = Relative percentage error between S_{G2} and S_{G3}. (b) $\varepsilon_{21}\%S_{G3}$ = Relative percentage error between S_{G2} and S_{G1}. (c) R_G = Grid convergence condition. (d) Con. = Convergence. (e) Mon. = Monotonic. (f) P_G = Estimated order of accuracy for grid convergence. (g) C_G = Correction factor of calculation for grid convergence. (h) $U_G\%S_{G3}$ = Discretization uncertainty of grid. (i) $U_I\%S_{G3}$ = Iterative uncertainty of grid.

Table 7. Results of time step convergence validation ($\lambda/L = 1.15$).

	$\varepsilon_{23}\%S_{T3}$ (a)	$\varepsilon_{21}\%S_{T3}$ (b)	R_T (c)	Con. (d)	P_T (f)	C_T (g)	$U_T\%S_{T3}$ (h)	$U_I\%S_{T3}$ (i)
K_Q	−0.1972	−0.1512	0.7669	Mon. (e)	0.3829	0.3040	0.6486	0.0107
K_T	−0.2774	−0.1323	0.4771	Mon.	1.0678	1.0962	0.0243	0.0006

(a) $\varepsilon_{23}\%S_{T3}$ = Relative percentage error between S_{T2} and S_{T3}. (b) $\varepsilon_{21}\%S_{T3}$ = Relative percentage error between S_{T2} and S_{T1}. (c) R_T = Time step convergence condition. (d) Con. = Convergence. (e) Mon. = Monotonic. (f) P_T = Estimated order of accuracy for time step convergence. (g) C_T = Correction factor of calculation for time step convergence. (h) $U_T\%S_{T3}$ = Discretization uncertainty of time step. (i) $U_I\%S_{T3}$ = Iterative uncertainty of time step.

Table 6 shows that both propeller thrust and torque coefficients converge monotonically with grid refinement. All convergence coefficients (C_G and C_T) are between 0 and 1, indicating favorable convergence characteristics of the calculation. The grid discretization uncertainty ($U_G\%S_{G3}$) indicates minimal sensitivity to grid size changes. Furthermore, the low iterative uncertainty ($U_I\%S_{G3}$) on the fine grid suggests iterative errors have minimal impact at this resolution. This collectively verifies grid convergence for the current simulations. Similarly, Table 7 demonstrates monotonic convergence of thrust and torque coefficients with time step refinement. The small-time step discretization uncertainty ($U_T\%S_{T3}$) confirms minimal sensitivity to time step size variations. The low iterative uncertainty ($U_I\%S_{T3}$) also shows iterative errors have minimal impact.

Additionally, the relative percentage error ($\varepsilon_{23}\%S_{G3}$) exhibits a small magnitude, indicating minimal differences between the results obtained with the S_{G2} and S_{G1} grids. Similarly, the relative percentage error $\varepsilon_{23}\%S_{T3}$ is also small, signifying minimal differences between the results using the S_{T2} and S_{T1} time steps. To achieve higher accuracy and computational efficiency in subsequent numerical simulations, the S_{G2} grid and S_{T2} time step will be employed accordingly.

4.3.2. Analysis of the Hydrodynamic Performance of the Propeller Considering the Influence of Heave and Pitch Motions at Different Wavelengths

Applying the proposed computational method, the hydrodynamic performances of the propeller were investigated under three different wavelengths are 1.00, 1.15 and 1.30. For each case, we configure the CFD simulation parameters according to the specifications provided in Tables 3 and 4. To better demonstrate the influence of heave and pitch motions on propeller performance, a supplementary simulation was conducted for each case, in which the same propeller rotational speed and advance ratio were maintained whereas the amplitudes of heave and pitch were set to zero.

Figure 9a displays the kinematic trajectories of propeller heave displacement and pitch angular motion under prescribed operational conditions at dimensionless wavelength is 1.15. Figure 9b presents the corresponding time histories of thrust and torque coefficients, comparing cases with and without heave and pitch motions of the propeller.

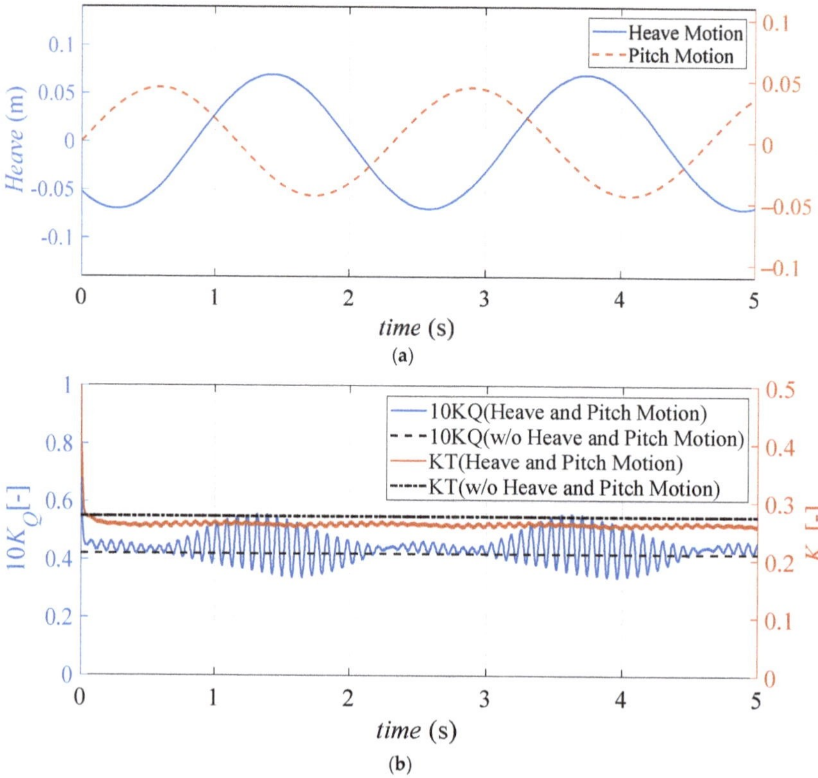

Figure 9. The motion state of the propeller and the variation of the hydrodynamic coefficient with time at λ/L = 1.15 (**a**) Kinematic profile of the propeller heave and pitch motions; (**b**) Computed torque and thrust coefficient.

As shown in Figure 9b, when propeller heave and pitch motions are taken into account, the time histories of thrust and torque coefficients exhibit two distinct periodic patterns. The shorter period corresponds to the blade frequency, while the longer period matches the frequency of heave and pitch motions (i.e., the wave encounter frequency). Compared to the results without heave and pitch motions, the inclusion of the motion leads to a decrease in the propeller's average thrust coefficient and an increase in the average torque coefficient.

Figure 10 presents the variations in thrust coefficient and torque coefficient for wavelengths dimensionless wavelength is 1.00 and dimensionless wavelength is 1.30; the observed trends of thrust reduction and torque increase are consistent with the conclusions drawn for the dimensionless wavelength is 1.15 case, exhibiting a decrease in mean thrust and an increase in mean torque. Numerically, the thrust coefficients for the propeller with heave and pitch motions decreased by approximately 5% across all wavelengths compared to the case without motions, while the torque coefficients increased by approximately 5%. This significant reduction has notable implications for ship propulsion performance. For instance, using only traditional open-water propeller curves to predict the self-propulsion point results in actual thrust being lower than expected values, preventing achievement of the target ship speed.

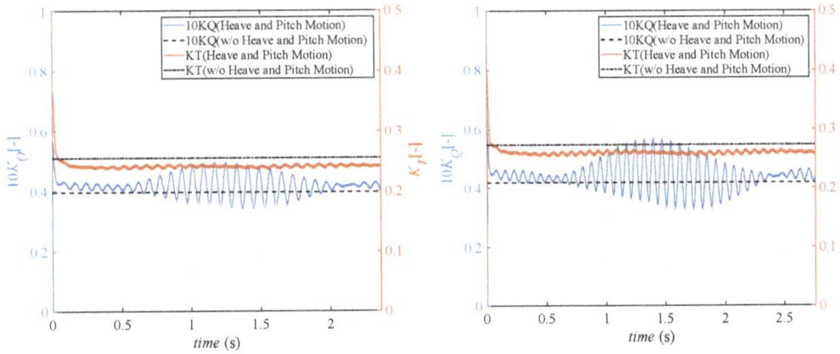

(a)

(b)

Figure 10. The variations of the propeller hydrodynamic coefficients at λ/L = 1.00 and λ/L = 1.30 (**a**) Wavelengths λ/L = 1.00; (**b**) Wavelengths λ/L = 1.30.

4.3.3. Validation of the Simplified Computational Method

To further validate the effectiveness of the proposed simplified algorithm, this study conducted additional CFD simulations of self-propelled ship-propeller systems in waves (which is referred to as the 'high-fidelity method' hereafter). The results from these high-fidelity simulations were then systematically compared with those obtained from the simplified approach. Figure 11 presents the schematic diagram and boundary condition configuration for the high-fidelity computations.

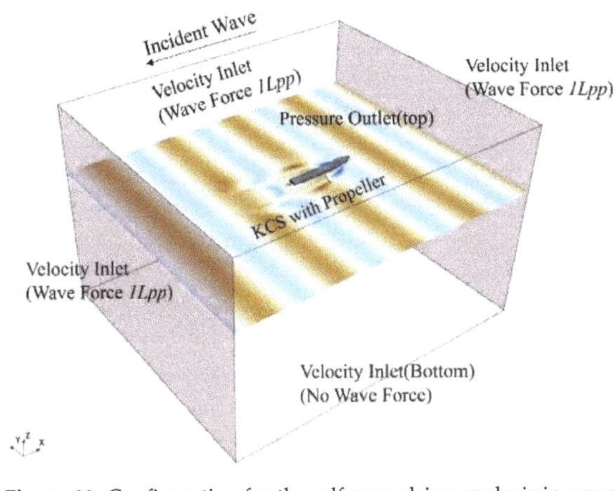

Figure 11. Configuration for the self-propulsion analysis in waves using a fully discretized ship–propeller system.

For the computational setup of the high-fidelity method, the DFBI approach was employed to simulate ship motions in waves. A PI controller was used to regulate the ship's speed to 2.196 m/s under self-propulsion conditions. The entire computational domain was defined as the background region, with an overset grid established around the hull. In cell size and position between the overset and background regions. Mesh morphing was activated to ensure a smooth and natural transition Furthermore, for the propeller, an overset grid was adopted to better capture the intricate flow characteristics. A schematic of the background and overset mesh is shown in Figure 12.

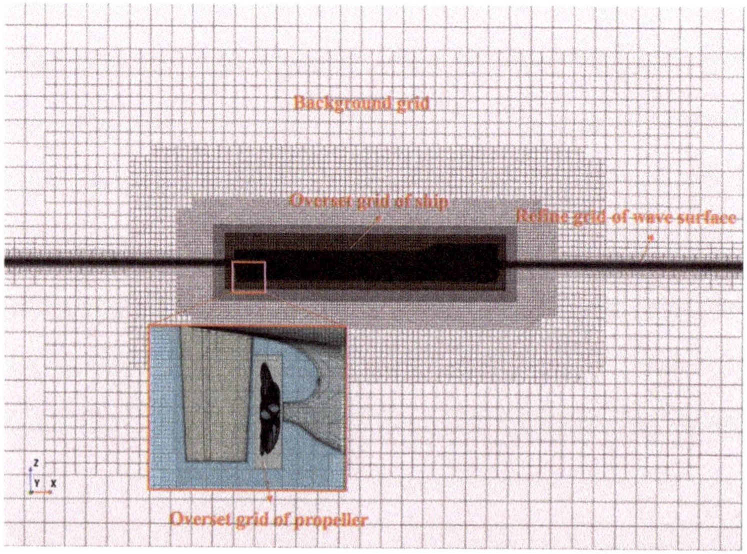

Figure 12. Mesh setup for the self-propulsion analysis in waves using a fully discretized ship-propeller system.

Furthermore, to capture wave characteristics and mitigate the influence of reflected waves, a wave-forcing method was employed. Wave-forcing zones, each with a length of one ship length, were applied on all sides of the computational domain. The mesh was discretized to ensure at least 80 cells per wavelength and 10 cells per wave height, with the aspect ratio of any individual cell not exceeding 8:8:1. A fifth-order VOF wave model was selected for wave generation. The time step was set to 2.07×10^{-4} s with a maximum of 10 internal iterations per step. The simulation was configured for a maximum physical time of 25 s.

Figure 13 presents the computed thrust and torque coefficients of the propeller. The red line with circle symbol represents results from the simplified algorithm accounting for heave and pitch motions, while the black dash line with square symbol shows results from the same algorithm without considering these motions. The blue dash-dot line with triangle symbol corresponds to the high-fidelity CFD results.

The results presented in Figure 13 confirm that heave and pitch motions significantly influence propeller performance, leading to thrust reduction and torque increment. As shown, the hydrodynamic parameters for the propeller without heave and pitch motions deviate substantially from the high-fidelity results. In comparison, the proposed simplified algorithm effectively narrows this discrepancy—particularly for the thrust coefficient—demonstrating its validity.

Although the simplified method accounting for heave and pitch motions shows improved results, some discrepancies with the high-fidelity data remain. These differences stem primarily from the method's assumption of uniform inflow (calculated using a wake fraction) rather than incorporating the actual non-uniform ship wake. Accurate modeling of non-uniform inflow, especially the complex wake field, remains computationally challenging. Furthermore, pressure fluctuations induced by ship motions in waves may significantly influence propeller cavitation dynamics. The underlying physics involved in these phenomena presents complex theoretical challenges that remain unresolved in the current research landscape.

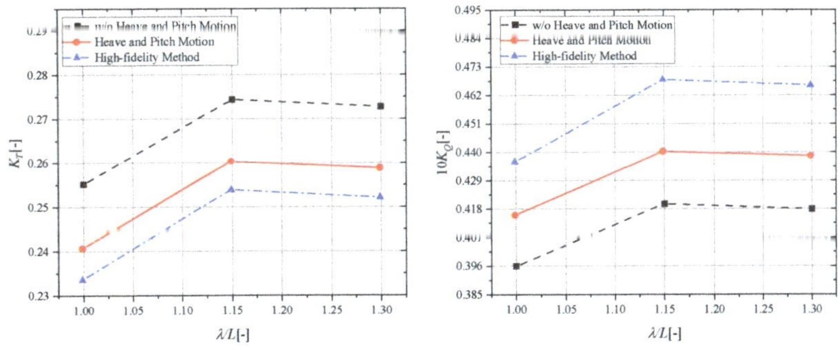

(**a**) (**b**)

Figure 13. Comparison of the hydrodynamic performance of propeller under different wavelengths in three design scenarios (**a**) Comparison of thrust coefficient; (**b**) Comparison of torque coefficient.

It is noteworthy that, under identical computational resources, the simplified algorithm is four times more efficient than the high-fidelity approach. This substantial improvement enables systematic investigation of propeller performance across various operating conditions. Given the significantly reduced computational burden, the accuracy attained by the proposed method—especially in predicting thrust coefficient—represents a practical and rational balance between computational expense and result reliability. Consequently, the proposed method shows strong potential as an efficient tool for the systematic analysis of propeller performance in complex operational scenarios.

4.3.4. Analysis on the Hydrodynamic Performance of the Propeller Considering Different Wave Steepness

Based on the simplified method, we further investigate wave height effects on propeller performance. To this end, three typical dimensionless wavelengths (the three dimensionless wavelengths are 1.00, 1.15, and 1.30) were selected and for each wavelength three wave steepness factors, 1/30, 1/60, and 1/90, were systematically examined.

Table 8 summarizes the wave steepness, advance coefficient and 'Ideal' propeller rotation for each computational case. The heave and pitch motions under each computational case, which were determined according to Table 3, were incorporated into the proposed method to evaluate the propeller's hydrodynamic performance across varying wave steepness conditions.

Table 8. Hull and propeller parameters used for the simulations at different wave steepness.

Dimensionless Wavelength λ/L	Wave Steepness H/λ	Advance Coefficient J	'Ideal' Propeller Rotation Speed n(rps)
1.00	1/30	0.432	16.091
	1/60	0.557	12.490
	1/90	0.598	11.643
1.15	1/30	0.375	18.553
	1/60	0.519	13.400
	1/90	0.575	12.101
1.30	1/30	0.378	18.411
	1/60	0.523	13.318
	1/90	0.577	12.075

Figure 14 compares thrust coefficients across different wave steepness values at constant wavelengths, while Figure 15 presents the corresponding torque coefficients. In both figures, results for the propeller under heave and pitch motions are indicated by red solid lines, whereas blue dotted lines represent the propeller without these motions.

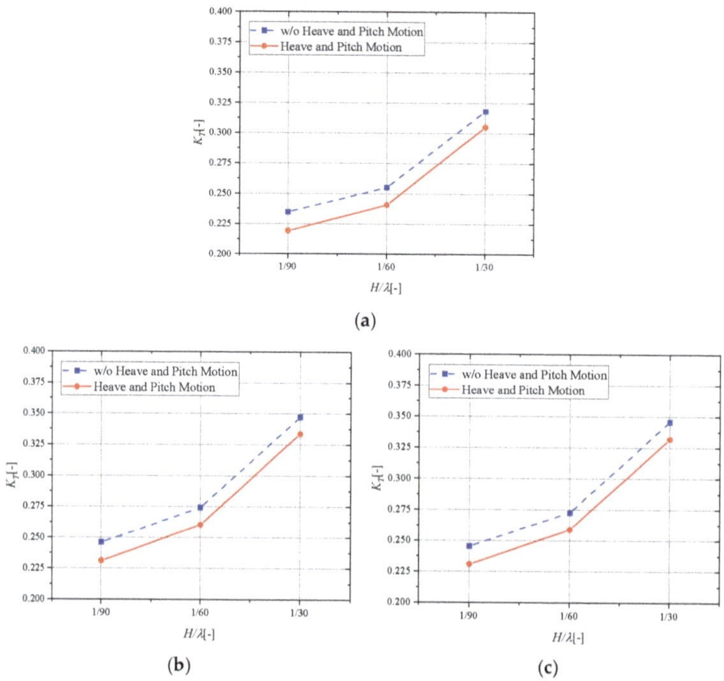

Figure 14. Comparison of thrust coefficients under different wave steepness of the same wavelength (**a**) Dimensionless wavelengths λ/L = 1.00; (**b**) Dimensionless wavelengths λ/L = 1.15; (**c**) Dimensionless wavelengths λ/L = 1.30.

As shown in Figures 14 and 15, both thrust and torque coefficients increase notably with greater wave steepness. This trend is attributed to the heightened wave-added resistance under steeper waves, which imposes an increased load on the propeller at a fixed sailing speed. Furthermore, the figures reaffirm the observed reduction in thrust and increase in torque resulting from heave and pitch motions.

Figure 16 illustrates the reduction in propeller efficiency caused by heave and pitch motions. This is examined for different wave steepness at a fixed wavelength. In the figure, the red solid line represents propeller efficiency incorporating heave and pitch effects, while the blue dotted line corresponds to the baseline efficiency without these motion-induced disturbances. The orange bar chart quantifies the absolute efficiency reduction between the two scenarios.

As seen in Figure 16, under the three tested wave steepness conditions, the propeller efficiency shows noticeable decreases when heave and pitch motions are considered compared to when they are not. Under low wave steepness of 1/90, the propeller efficiency experiences an absolute decrease of over six percentage points when accounting for heave and pitch effects, which is equivalent to a relative reduction exceeding 10%; whereas under high wave steepness of 1/30, the absolute efficiency drop exceeds three percentage points, corresponding to a relative reduction of approximately 7%.

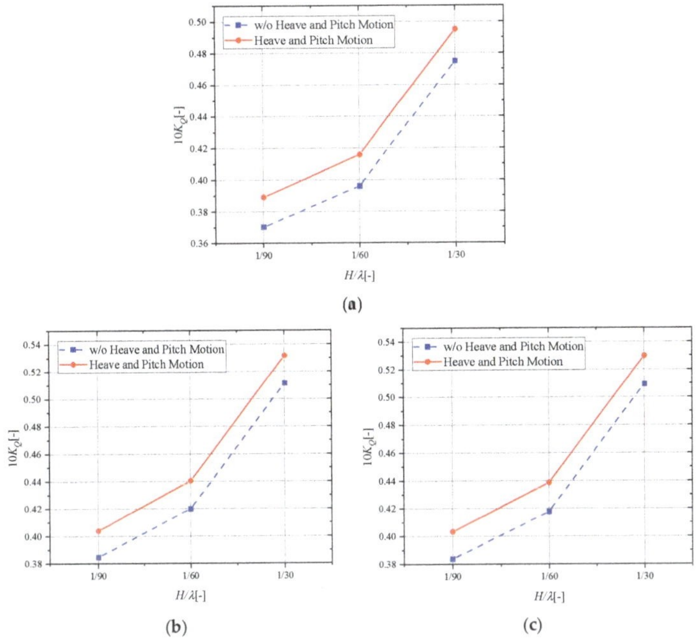

Figure 15. Comparison of torque coefficients under different wave steepness of the same wavelength (**a**) Dimensionless wavelengths λ/L = 1.00; (**b**) Dimensionless wavelengths λ/L = 1.15; (**c**) Dimensionless wavelengths λ/L = 1.30.

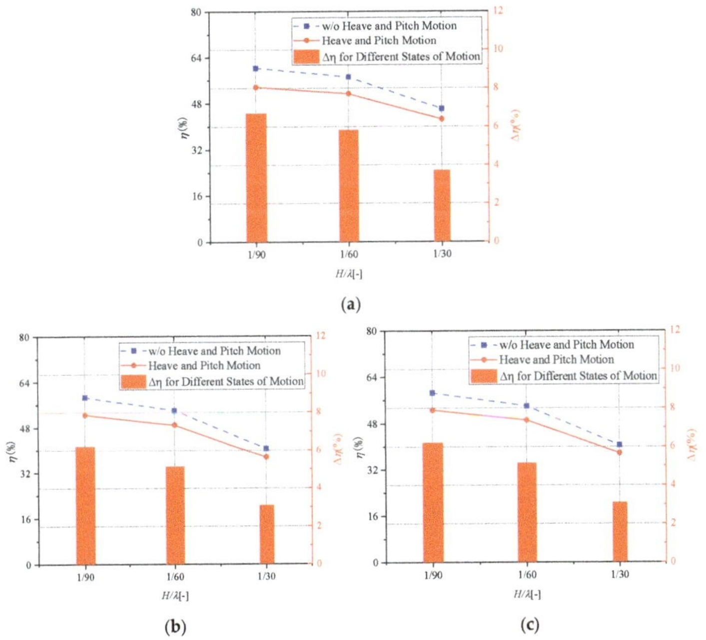

Figure 16. Comparison of propeller efficiencies under different wave steepness of the same wavelength (the orange bars indicate the absolute efficiency reduction of the propeller caused by heave and pitch motions) (**a**) Dimensionless wavelengths λ/L = 1.00; (**b**) Dimensionless wavelengths λ/L = 1.15; (**c**) Dimensionless wavelengths λ/L = 1.30.

It is interesting to find that the influence of heave and pitch motions on propeller efficiency is less pronounced under high wave steepness compared to low wave steepness conditions. This phenomenon can be attributed to the following mechanisms: while larger wave steepness corresponds to greater heave and pitch motion amplitudes, it simultaneously induces more significant wave-added resistance, consequently altering the propeller's operating conditions under self-propulsion. At lower wave steepness, the propeller operates at a relatively lower rotational rate and, therefore, the circumferential inflow velocity of the propeller is relatively low. When the radial inflow velocity of the propeller is affected by heave and pitch motions, the resulting changes in the hydrodynamic pitch angle of propeller blades become more pronounced, leading to greater variations in overall hydrodynamic performance. Conversely, at higher wave steepness, the propeller rotates faster, and the circumferential inflow velocity increases. In this scenario, the influence of heave and pitch motions on the advance velocity has relatively less impact on the propeller's overall performance, despite the larger heave and pitch motion amplitudes.

4.3.5. Impact of Independent Heave and Pitch Motions on Propeller Hydrodynamics

In all the numerical results presented above, both heave motion and pitch motion were simultaneously applied to the propeller during calculations. To better understand their individual contributions to propeller hydrodynamics, this section separately examines the effects of isolated heave motion and pitch motion. Following the same operating conditions listed in Tables 3 and 4, we applied these individual motions to our proposed simplified method.

Figure 17 presents variations in propeller hydrodynamic performance across varying wavelength conditions, under three kinematic configurations: coupled heave and pitch motions (black dotted line with square symbol), isolated heave motion (blue dotted line with circle symbol), and isolated pitch motion (red dash-dotted line with triangle symbol).

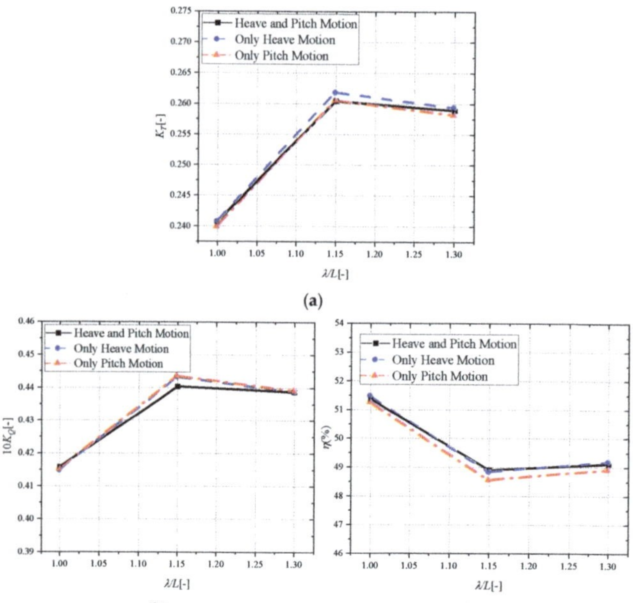

Figure 17. Comparative analysis of hydrodynamic performance across varying wavelength for different propeller motion configurations (**a**) Comparison of thrust coefficient; (**b**) Comparison of torque coefficient; (**c**) Comparison of propeller efficiency.

Figure 17a shows that isolated heave motion generates higher thrust than coupled heave-pitch motion, while isolated pitch motion results in lower thrust. Figure 17b further indicates that both isolated heave and pitch motions produce greater torque than the coupled motion. Interestingly, Figure 17c reveals that the coupled heave-pitch motion does not result in the lowest efficiency among the three cases—instead, the isolated pitch motion leads to the most substantial efficiency reduction. These findings suggest that hydrodynamic coupling between heave and pitch motions introduces a nonlinear, counteracting effect on efficiency loss, rather than a simple additive combination of individual motion penalties. As can be seen from the figure, compared to heave motion, the ship's pitch motion has a more significant impact on propeller performance, not only reducing the thrust but also causing a noticeable decrease in propeller efficiency.

5. Conclusions

This study develops a simplified computational method for assessing propeller performance under wave-induced heave and pitch motions. The methodology integrates potential flow theory-derived motion RAOs with CFD-based hydrodynamic analysis, where prescribed heave and pitch motions are systematically imposed on the propeller.

Numerical tests were conducted based on the KCS container ship and its matched KP505 propeller. The numerical results demonstrated that the wave-induced heave and pitch motions lead to a consistent performance degradation, manifested by thrust reduction and torque increase, ultimately leading to an absolute efficiency decrease of 3 to 6 percentage points across all tested wave conditions (equivalent to approximately 10% relative reduction).

Verification against high-fidelity CFD results confirms that the proposed method delivers fairly good accuracy and high computational efficiency. It therefore shows strong potential as an effective tool for the systematic analysis of propeller performance in complex operational scenarios.

Author Contributions: Conceptualization and methodology, W.Z. and R.W.; investigation and validation, G.X. and S.W.; formal analysis, G.X. and W.Z.; resources, R.W.; data curation, G.X.; writing—original draft preparation, G.X.; writing—review and editing, W.Z.; visualization, G.X. and S.W.; supervision, R.W.; project administration, W.Z. and R.W. All authors have read and agreed to the published version of the manuscript.

Funding: This research was funded by Open Funding Project of the State Key Laboratory of Maritime Technology and Safety.

Data Availability Statement: Data will be made available on request to the corresponding author.

Conflicts of Interest: Author Rui Wu was employed by Shanghai Ship and Shipping Research Institute Co., Ltd. The remaining authors declare that the research was conducted in the absence of any commercial or financial relationships that could be construed as a potential conflict of interest.

References

1. IMO. *Guidelines on the Method of Calculation of the Attained Energy Efficiency Design Index (EEDI) for New Ships*; International Maritime Organization: London, UK, 2018.
2. IMO. *EEXI and CII—Ship Carbon Intensity and Rating System*; IMO International Maritime Law Institute: Msida, Malta, 2021.
3. Joung, T.-H.; Kang, S.-G.; Lee, J.-K.; Ahn, J. The IMO initial strategy for reducing Greenhouse Gas(GHG) emissions, and its follow-up actions towards 2050. *J. Int. Marit. Saf. Environ. Aff. Shipp.* **2020**, *4*, 1–7. [CrossRef]
4. Tavakoli, S.; Saettone, S.; Steen, S.; Andersen, P.; Schramm, J.; Pedersen, E. Modeling and analysis of performance and emissions of marine lean-burn natural gas engine propulsion in waves. *Appl. Energy* **2020**, *279*, 115904. [CrossRef]
5. Saettone, S.; Tavakoli, S.; Taskar, B.; Jensen, M.V.; Pedersen, E.; Schramm, J.; Steen, S.; Andersen, P. The importance of the engine-propeller model accuracy on the performance prediction of a marine propulsion system in the presence of waves. *Appl. Ocean Res.* **2020**, *103*, 102320. [CrossRef]

6. Irannezhad, M.; Kjellberg, M.; Bensow, R.E.; Eslamdoost, A. Experimental and numerical investigations of propeller open water characteristics in calm water and regular head waves. *Ocean Eng.* **2024**, *302*, 117703. [CrossRef]
7. Acanfora, M.; De Luca, F.; Pigazzini, R.; Altosole, M. An experimental study on the open water characteristics of a ship propeller in waves. *Appl. Ocean Res.* **2025**, *154*, 104331. [CrossRef]
8. Cosgun, T.; Esenkalan, M.; Kinaci, O.K. Four-quadrant propeller hydrodynamic performance mapping for improving ship motion predictions. *Brodogradnja* **2024**, *75*, 306. [CrossRef]
9. Liang, L.; Hao, Z.; Chaonan, Z.; Jinbiao, C.; Baoji, Z.; Xiangen, B.; Shengyao, S.; Qian, C.; Weijia, Z. Study on the effect of marine propeller wake on sediment siltation in a shallow water channel. *Brodogradnja* **2024**, *75*, 75308–75322. [CrossRef]
10. Yurtseven, A.; Aktay, K. The numerical investigation of spindle torque for a controllable pitch propeller in feathering maneuver. *Brodogradnja* **2023**, *74*, 95–108. [CrossRef]
11. Grlj, C.G.; Degiuli, N.; Farkas, A.; Martić, I. Numerical Study of Scale Effects on Open Water Propeller Performance. *J. Mar. Sci. Eng.* **2022**, *10*, 1132. [CrossRef]
12. Ueno, M.; Tsukada, Y.; Tanizawa, K. Estimation and prediction of effective inflow velocity to propeller in waves. *J. Mar. Sci. Technol.* **2013**, *18*, 339–348. [CrossRef]
13. Ueno, M.; Tsukada, Y. Estimation of full-scale propeller torque and thrust using free-running model ship in waves. *Ocean Eng.* **2016**, *120*, 30–39. [CrossRef]
14. Cai, B.; Qiu, L.; Tian, B.; Xu, Q.; Mao, X.; Chai, W.; Zhan, X. Research on predicting methods of propeller-hull interactions in head waves. *Ocean Eng.* **2023**, *269*, 113493. [CrossRef]
15. Gamal, H.; Kotb, M.A.; Abdel-Maksoud, M.; Ahmed, T.M. Influence of Head Waves Encounter on the KRISO Container Ship (KCS) Propulsion. In Proceedings of the 8th International Symposium on Marine Propulsors (smp 2024), Berlin, Germany, 17–20 March 2024. [CrossRef]
16. Woeste, J.T.; O'reilly, C.M.; Gouveia, R.K.; Young, Y.L. Propeller–hull interactions and added power in head seas. *Ocean Eng.* **2022**, *247*, 110630. [CrossRef]
17. Sadat-Hosseini, H.; Toxopeus, S.; Kim, D.H.; Castiglione, T.; Sanada, Y.; Stocker, M.; Simonsen, C.; Otzen, J.F.; Toda, Y.; Stern, F. Experiments and Computations for KCS Added Resistance for Variable Heading. In Proceedings of the SNAME Maritime Convention and 5th World Maritime Technology Conference, Providence, RI, USA, 3–7 November 2015.
18. Ortolani, F.; Santic, I.; Dubbioso, G. On the performance of marine propeller in regular waves, Part I: Single Blade loads in moderate sea states. *Ocean Eng.* **2024**, *309*, 118290. [CrossRef]
19. Wang, L.; Guo, C.; Su, Y.; Xu, P.; Wu, T. Numerical analysis of a propeller during heave motion in cavitating flow. *Appl. Ocean Res.* **2017**, *66*, 131–145. [CrossRef]
20. Wang, L.; Guo, C.; Wan, L.; Su, Y.M. Numerical analysis of propeller during heave motion near a free surface. *Mar. Technol. Soc. J.* **2017**, *51*, 40–51. [CrossRef]
21. Lee, J.-H.; Kim, Y. Study on ship operation performance in actual seaways using time-domain free-running simulation. *Proc. Inst. Mech. Eng. Part M J. Eng. Marit. Environ.* **2020**, *235*, 885–904. [CrossRef]
22. Okuda, R.; Yasukawa, H.; Matsuda, A. Validation of maneuvering simulations for a KCS at different forward speeds using the 4-DOF MMG method. *Ocean Eng.* **2023**, *284*, 115174. [CrossRef]
23. Stark, C.; Shi, W. Hydroacoustic and hydrodynamic investigation of bio-inspired leading-edge tubercles on marine-ducted thrusters. *R. Soc. Open Sci.* **2021**, *8*, 210402. [CrossRef] [PubMed]
24. Lee, S.; Paik, K.-J. Urans simulation of a partially submerged propeller operating under the bollard condition. *Brodogradnja* **2018**, *69*, 107–121. [CrossRef]
25. Mikkelsen, H.; Shao, Y.; Walther, J.H.H. CFD verification and validation of added resistance and seakeeping response in regular oblique waves with varying wave length. In Proceedings of the 9th Conference on Computational Methods in Marine Engineering (Marine 2021), Edinburgh, UK, 2–4 June 2021. [CrossRef]
26. Song, S.; Demirel, Y.K.; Atlar, M. Penalty of hull and propeller fouling on ship self-propulsion performance. *Appl. Ocean Res.* **2020**, *94*, 102006. [CrossRef]
27. Stern, F.; Wilson, R.V.; Coleman, H.V.; Paterson, E.G. Verification and validation of CFD simulations. In Proceedings of the 3rd ASME/JSME Joint Fluids Engineering Conference: FEDSM'99, San Francisco, CA, USA, 18–23 July 1999.
28. Grlj, C.G.; Degiuli, N.; Martić, I. Experimental and numerical assessment of the effect of speed and loading conditions on the nominal wake of a containership. *Brodogradnja* **2024**, *75*, 75405. [CrossRef]

Journal of
Marine Science and Engineering

MDPI

Article

Comparison of Numerical Simulations of Propeller Open-Water Performance with Cavitation for High-Speed Planing Hulls

Sungtek Park [1], Zhaoyuan Wang [1,*], Frederick Stern [1,*], Andrew Gunderson [2] and John Scherer [2]

1 IIHR-Hydroscience & Engineering, University of Iowa, Iowa City, IA 52242, USA; sungtaek-park@uiowa.edu
2 Mercury Marine, Fond Du Lac, WI 54935, USA; john.scherer@mercmarine.com (J.S.)
* Correspondence: zhaoyuan-wang@uiowa.edu (Z.W.); frederick-stern@uiowa.edu (F.S.)

Abstract: Numerical simulations of an open-water propeller are performed using CFDShip-Iowa. The propeller, originally designed by Mercury Marine for a 21 feet high-speed planing hull, is scaled to match a 42 feet hull configuration. Three advance ratios (J = 0.8, 1.1, and 1.4) and two cavitation numbers (σ = 0.274 and 1.095) are considered in the computations, and the results are compared with those obtained from the commercial CFD solver STAR-CCM+. For the fully wetted conditions without cavitation, the overall trends of the computed thrust (K_t), torque (K_q), and propeller efficiency (η) with respect to the advance ratios are similar. The computed K_t, K_q, and η with cavitations generally agree with the STAR-CCM+ results except for η at σ = 0.274, where the latter shows a much higher value for J = 1.4. For σ = 1.095, the cavitation patterns and overall pressure distributions are similar for both codes. For σ = 0.274, the cavitation is more violent for CFDShip-Iowa than STAR-CCM+. CFDShip-Iowa shows better preservation of the cavities and blade-to-blade interactions, which are not captured in the simulations using STAR-CCM+, since a single blade with periodic boundary conditions are used.

Keywords: propeller open-water; CFD; high-speed planning hull; cavitation

1. Introduction

High-speed small craft with surface-pierced propellers experience severe cavitation and ventilation effects, particularly during planing conditions at high speeds. Ventilation occurs when air is drawn into the propeller region due to partial submergence, leading to a decrease in thrust and overall propeller efficiency. Cavitation induces noise, vibration, and potential damage to propeller blades, negatively affecting vessel performance and longevity. Accurately capturing the interaction of these multiphase phenomena with the rotating propeller is essential for reliable performance prediction and design optimization.

Propeller open-water (POW) tests with cavitation effects play a critical role in estimating the performance of high-speed planing hulls. However, both experimental testing and numerical prediction of propeller hydrodynamics with cavitation phenomena is challenging due to the interactions between the complex turbulent multiphase flows and the rotating body. Experimental POW tests involving cavitation are more limited by scale effects, facility constraints, controllability of test conditions, and difficulties in visualization and measurement. An experimental study for a surface-piercing propeller was conducted by Olofsson [1] using the propeller model 841-B, followed by the numerical studies by Young and Kinnas [2] and Nouroozi and Zeraatgar [3] using a boundary element method (BEM) and a (Unsteady Reynolds-Averaged Navier–Stokes) URANS method, respectively. The BEM results generally compared well with the experimental measurements, but force fluctuations were not captured. The URANS results computed using the FLUENT (v14.5) software showed errors up to 15%. In the recent study by Yari and Moghadam [4], the POW characteristics of the 841-B propeller were studied using BEM-based predictions. The computed results showed small errors at high advance ratios but large errors at low advance ratios, where the errors for the thrust and torque coefficients were up to 30% and 60%, respectively, at an advance ratio of 0.4. Therefore, high-fidelity Computational Fluid Dynamics (CFD) simulations that incorporate viscosity effects should be considered for accurately predicting a wide range of propeller operating conditions. Numerous CFD POW studies have been conducted recently to investigate cavitation effects [5–9], salinity effects [10], wave conditions [11], geometry optimization [12], scaling effects [13], and propeller–hull interaction [14], though most studies are not specifically for propellers on high-speed planing hulls. Overall, these studies demonstrate that CFD approaches are capable and effective for evaluating marine propeller performance under various conditions.

The objective of the present study is to perform numerical simulations of the propeller open-water performance with a focus on cavitation effects for a high-speed planing hull, which serves as a prerequisite for subsequent self-propulsion and acceleration simulations [15] using the actual propeller and gear case geometries. A Froude number of 1.84 is used for the simulations, which is relatively high compared to those usually employed in POW tests. The PLSN6143 propeller, originally designed by Mercury Marine for a 21 feet (6.4 m) high-speed planing hull, is scaled up by a factor of two to match a 42 feet (12.8 m) model of General Prismatic Planing Hull (GPPH) and is used for the simulations. The high-fidelity simulation tools CFDShip-Iowa V4.5 and V5.5 are employed for the computations, with the results compared to those obtained from the commercial CFD solver STAR-CCM+. Detailed computational methods and the setup are presented in Section 2, followed by the computational results and discussions in Section 3.

2. Computational Methods and Setup

CFDShip-Iowa V4.5 and V5.5 are used for the simulations. V4.5 is a single-phase solver with the level set method for free surface modeling, and V5.5 is a fully coupled multi-phase solver [16,17] using a volume-of-fluid (VOF) method for the air–water interface treatment. The grids and CFD setups are the same for both codes. V4.5 is used for the fully wetted cases without cavitation, and V5.5 is used for both fully wetted and cavitation cases. The commercial CFD code, STAR-CCM+ v17.02, is also used for the simulations for comparison purposes. Note that the use of two distinct solvers is for the independent verification of the results since experimental data is not available. The choice of turbulence and cavitation models and overall modeling strategies is based on model availability, code stability, and practical experience.

2.1. Mathematical Model and Numerical Method

In CFDShip-Iowa V4.5, a single-phase level set method is used for free surface modeling. The pressure in the air region is constant, and a zero normal gradient for both the k and ω are used at the free surface. Periodic reinitializations of the level set function are performed to keep the property of being a distance function during the simulation. CFDShip-Iowa V5.5 is a fully coupled multi-phase flow solver developed based on V4.5. In V5.5, the highly accurate geometric VOF method is used to replace the level set method for interface tracking. The vapor transport mixture cavitation models, e.g., the Zwart–Gerber–Belamri model [18], have been implemented into V5.5 [17], where a numerical scheme has been developed for the cavitation models in the framework of a pressure projection method for incompressible Navier–Stokes equations. An algorithm for three-phase water/vapor/air interaction is also developed for flows with both ventilation and cavitation. In the present study, cavitation is represented using the vapor volume fraction.

Menter's blended $k - \omega/k - \epsilon$ turbulent model [19] is used for turbulence modeling in both V4.5 and V5.5. The 6DoF motion of ships is predicted by solving the rigid body equations of motion and is realized using the overset grid technique. Suggar [20] is used to obtain the overset domain connectivity information for overlapping grids. An implicit second-order Euler backward difference scheme is used for the time derivatives. The convection terms are discretized using the 2nd-order upwind scheme, and the 2nd-order central difference scheme is used for the diffusion terms. The same discretization schemes are used for the turbulence and level set equations. Incompressibility is enforced by a strong pressure/velocity coupling using a projection (fractional time step) method, and the resulting pressure equation is solved using the Portable Extensible Toolkit for Scientific Computing (PETSc) v3.19.4. The code is parallelized using the MPI-based domain decomposition approach, where each decomposed block is mapped to one processor. Details of the mathematical models and numerical methods can be found in the above papers and references therein.

STAR-CCM+ (v17.02) employs unstructured polyhedral grids and the finite volume method for spatial discretization, along with an implicit scheme for time integration. The second-order convective HRIC scheme volume-of-fluid (VoF) method is used to capture and track the water–vapor interface. The realizable $k - \varepsilon$ two-layer model is used for the turbulence modeling, and the Schnerr–Sauer cavitation model [21] is used to model the phase change.

2.2. Computational Setup

Figure 1 shows the geometry of the PLSN6143 propeller, which has a diameter of 0.710 m and four blades at a skew angle of 11.77 degrees. The tip rake and skew-induced rake are 0.174 m and 0.035 m, respectively. The pitch distance is 1.065 m, and the hub diameter is 0.228 m, which gives a hub-to-diameter ratio greater than 0.32. The length of the hub shown in Figure 1 is 0.37 m. The detailed specifications of the propeller geometry are given in Table 1. Note that both the STAR-CCM+ and CFDShip-Iowa simulations assume an infinite hub, with the latter extending to the inlet boundary. Only the blades are considered for the calculation of the thrust and torque coefficients.

(a) (b) (c)

Figure 1. Geometry of the PLSN6143 propeller. (**a**) Side view, (**b**) front view, (**c**) back view.

Figure 2 shows the computational domain in terms of propeller diameter (D) and boundary conditions for the CFDShip-Iowa simulations, where the hub of the propeller is extended to the inlet and a cylindrical background grid is used. Zero-gradient boundary conditions for both velocity and pressure are applied on the outer boundary of the cylindrical background grid. Constant velocity and zero gradient of the pressure are used at the inlet, and constant pressure and zero gradient of the velocity are imposed at the outlet. No-slip wall boundary conditions are applied to surfaces of the propeller blades and hub. The entire grid system rotates to represent the propeller rotating motion while maintaining the static overset mesh connectivity among grid blocks.

Table 1. Detailed specifications of PLSN6143 propeller.

Number of Blades	Propeller Diameter	Hub Diameter	Pitch Distance
4	0.71 m	0.2282 m	1.065 m
Skew	Tip rake	Skew-induced rake	Disc area
11.77 deg	0.1739 m	0.0348 m	0.3959 m^2

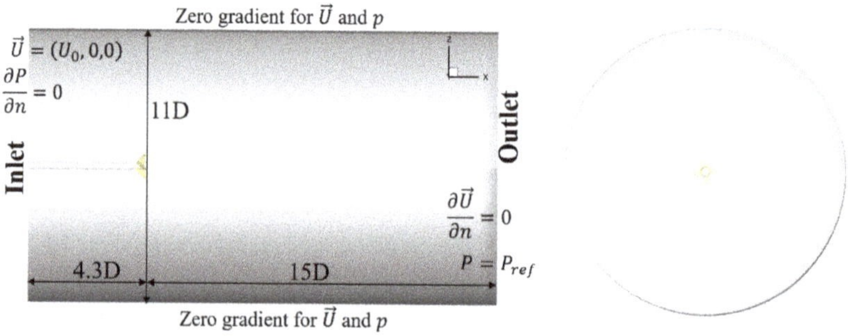

(**a**) (**b**)

Figure 2. Computational domain and boundary conditions for the CFDShip-Iowa simulations. (**a**) View from side; (**b**) view from the outlet.

Figure 3 shows the multi-block and structured overset grids used in the CFDShip-Iowa simulations. The grid consists of 10 overlapping blocks, including 1 background, 1 refinement, 4 blade, and 4 blade tip blocks. The numbers of grid points are 7.43 M, 3.38 M, 3.43 M, and 1.30 M, respectively, with a total of 15.5 M grid points. The grid spacing near the wall surface is designed to satisfy $y+ \approx 1.0$.

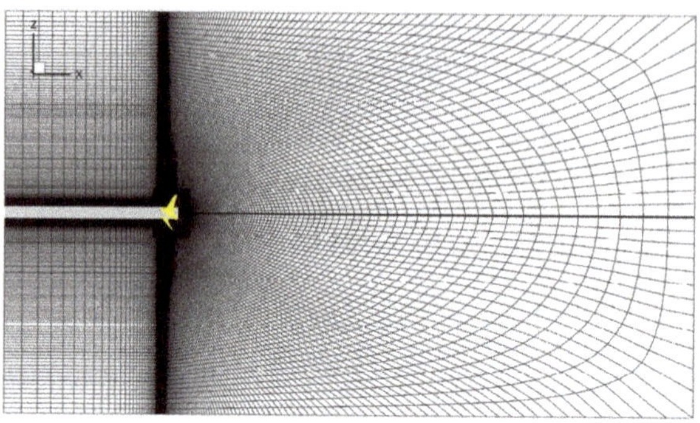

(**a**)

Figure 3. *Cont.*

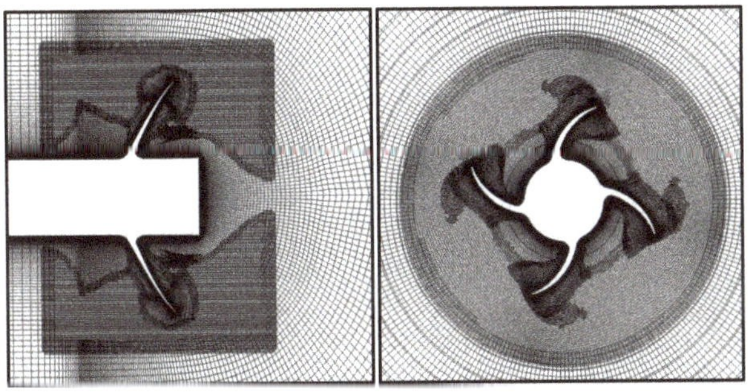

(b)

Figure 3. Grid distribution for the CFDShip-Iowa. (**a**) Background grid distribution; (**b**) propeller blade grid with the overset grid distribution.

For the STAR-CCM+ simulations, a single blade geometry is used. Figure 4 shows the geometry and computational domain, where a moving reference frame (MRF) and periodic boundary condition are used. Constant-velocity and zero-pressure-gradient boundary conditions are used at the inlet, and a constant-pressure boundary condition is applied at the outlet. The reference pressure P_{ref} is calculated based on the cavitation number:

$$\sigma = \frac{P_{ref} - P_v}{\frac{1}{2}\rho_w U_0^2} \quad (1)$$

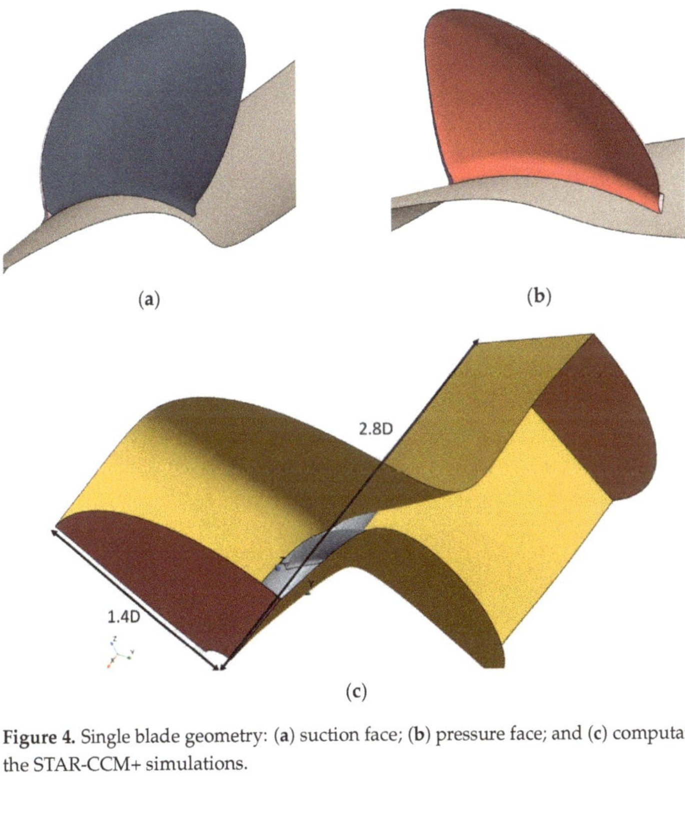

(a) (b) (c)

Figure 4. Single blade geometry: (**a**) suction face; (**b**) pressure face; and (**c**) computational domain for the STAR-CCM+ simulations.

An unstructured polyhedral mesh is used for the STAR-CCM+ simulations, as shown in Figure 5. The base size is set to 0.01 m, with target and minimum surface sizes set to 25% and 1% of the base size, respectively. At the no-slip wall, 5 prism layers are applied with a total thickness of 20% base size and a stretching factor of 1.5 to resolve the boundary layer appropriately. The total number of cells, faces, and vertices for the unstructured polyhedral mesh is 717 K, 3.78 M, and 2.81 M, respectively.

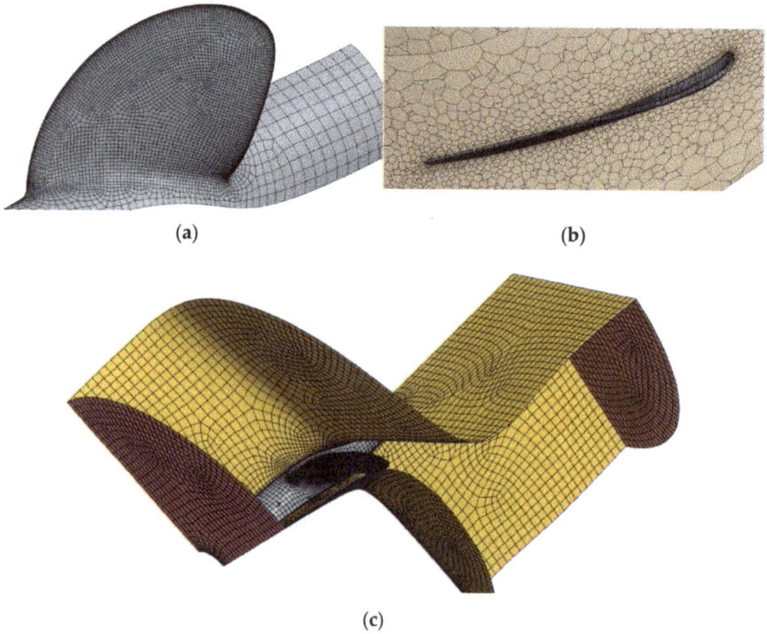

Figure 5. Grid distribution for the STAR-CCM+ simulations. (**a**) Single blade grid distribution; (**b**) polyhedral volume grid with prism layers; (**c**) background grid distribution.

For the current propeller open-water simulations, three advance ratios, $J = 0.8$, 1.1, and 1.4, are used with a constant inlet velocity of 20.68 m/s, which corresponds to a Froude number of $Fr = 1.84$ for the GPPH. The corresponding RPS values are 36.42, 26.49, and 20.81 for the given advance ratios. For the fluid properties, the densities $\rho_w = 1000$ kg/m^3 and $\rho_v = 0.02558$ kg/m^3 are used for the liquid and vapor phases, respectively. The kinematic viscosities are $\nu_w = 1.0 \times 10^{-6}$ m^2/s for the liquid phase and $\nu_v = 8.85 \times 10^{-9}$ m^2/s for the vapor phase. The vapor pressure for cavitation is $P_v = 2334.1$ Pa. Two different cavitation numbers, $\sigma = 0.274$ and 1.095, are considered, which require the reference pressure P_{ref} to be set to 60.94 kPa and 236.6 kPa, respectively.

3. Results

3.1. Thrust and Torque Coefficients and Propeller Efficiency

The computed results of the thrust, torque, and propeller efficiency using both CFDShip-Iowa and STAR-CCM+ are presented in Figures 6–8 and included in Tables 2–4.

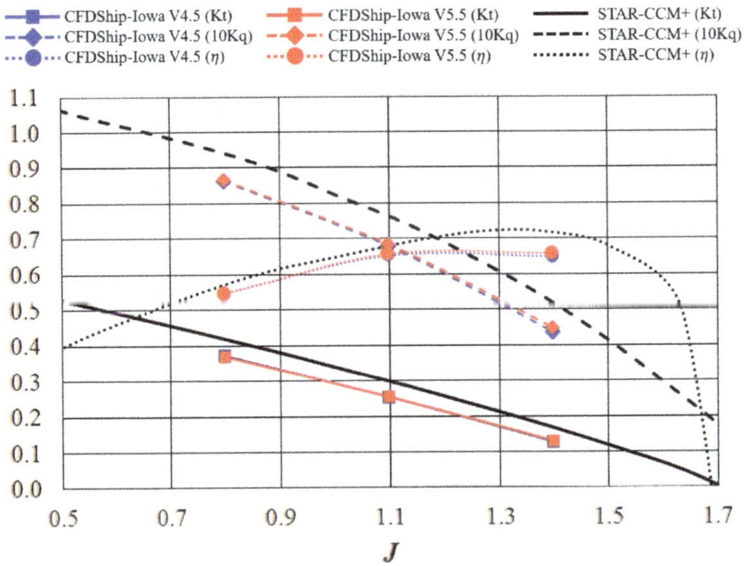

Figure 6. Thrust and torque coefficients and propeller open-water efficiency without cavitation.

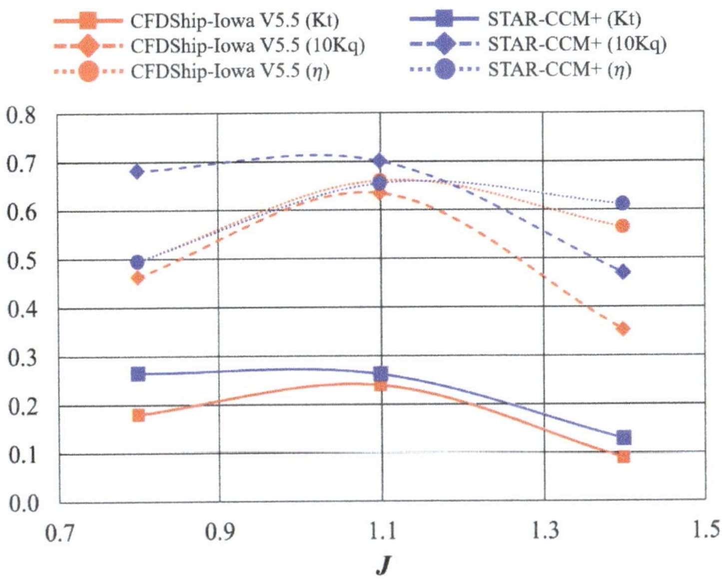

Figure 7. Thrust and torque coefficients and propeller open-water efficiency for the cavitation number of $\sigma = 1.095$.

Figure 6 presents the thrust and torque coefficients along with the propeller efficiency for the non-cavitating conditions. For the CFDShip-Iowa simulations, both the single-phase flow solver V4.5 and multi-phase flow solver V5.5 are used. As shown in the figure, the simulation results of both V4.5 and V5.5 are almost identical. For the STAR-CCM+ simulations, 14 advance ratios ranging from $J = 0.126$ to $J = 1.763$ are used. Generally, the results computed using STAR-CCM+ and CFDShip-Iowa show similar trends and are in good agreement. For the J values from 0.8 to 1.4, the thrust and torque coefficients decrease, and the propeller efficiency increases with the advance ratios. The thrust and torque values obtained using STAR-CCM+ are slightly higher than those obtained using CFDShip-Iowa. The propeller efficiency of STAR-CCM+ is also higher compared to CFDShip-Iowa, especially at the large advance ratio $J = 1.4$. As mentioned in the previous section, only the blades are considered for the calculation of the thrust and torque coefficients for CFDShip-Iowa.

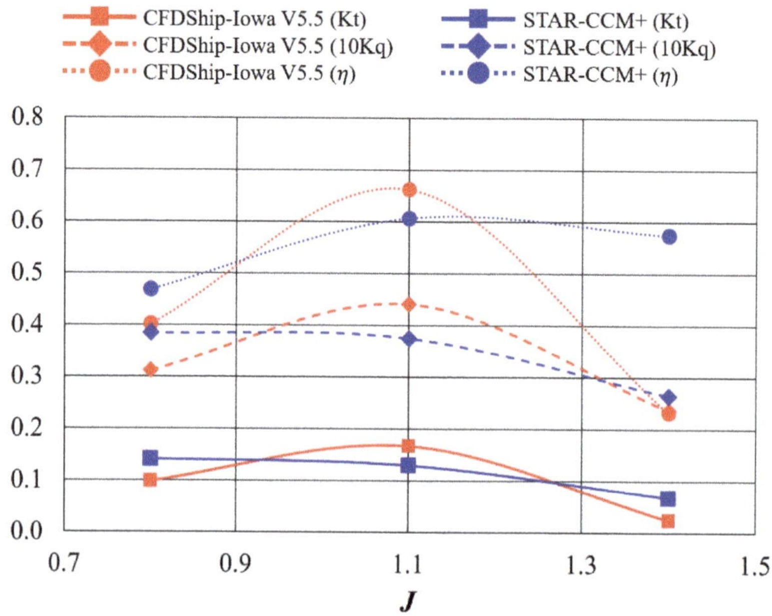

Figure 8. Thrust and torque coefficients and propeller open-water efficiency for the cavitation number of $\sigma = 0.274$.

Table 2. Thrust coefficients for CFDShip-Iowa V5.5 and STAR-CCM+.

Kt	**CFDShip-Iowa V5.5**					**STAR-CCM+**				
σ	No-Cav	1.095	Diff (%)	0.274	Diff (%)	No-Cav	1.095	Diff (%)	0.274	Diff (%)
$J = 0.8$	0.372	0.181	−51.4	0.098	−73.5	0.418	0.266	−36.5	0.142	−66.2
$J = 1.1$	0.257	0.240	−6.7	0.167	−35.1	0.298	0.263	−11.7	0.130	−56.4
$J = 1.4$	0.133	0.090	−32.4	0.025	−81.4	0.166	0.129	−21.9	0.068	−58.9
Average			−30.2		−63.4			−23.4		−60.5

Table 3. Torque coefficient for CFDShip-Iowa V5.5 and STAR-CCM+.

10Kq	**CFDShip-Iowa V5.5**					**STAR-CCM+**				
σ	No-Cav	1.095	Diff (%)	0.274	Diff (%)	No-Cav	1.095	Diff (%)	0.274	Diff (%)
$J = 0.8$	0.867	0.464	−46.5	0.312	−64.1	0.939	0.683	−27.3	0.385	−59.0
$J = 1.1$	0.685	0.635	−7.3	0.441	−35.6	0.761	0.702	−7.8	0.375	−50.8
$J = 1.4$	0.449	0.353	−21.3	0.236	−47.4	0.515	0.471	−8.4	0.264	−48.6
Average			−25.0		−49.0			−14.5		−52.8

Table 4. Propeller open-water efficiency for CFDShip-Iowa V5.5 and STAR-CCM+.

η	CFDShip-Iowa V5.5					STAR-CCM+				
σ	No-Cav	1.095	Diff (%)	0.274	Diff (%)	No-Cav	1.095	Diff (%)	0.274	Diff (%)
$J = 0.8$	0.546	0.496	−9.2	0.402	−26.4	0.571	0.496	−13.2	0.469	−18.0
$J = 1.1$	0.657	0.661	0.7	0.662	0.8	0.680	0.655	−3.5	0.606	−10.8
$J = 1.4$	0.658	0.565	−14.1	0.232	−64.8	0.714	0.612	−14.3	0.574	−19.6
Average			−7.6		−30.1			−10.4		−16.1

Figure 7 shows the simulation results for the high cavitation number $\sigma = 1.095$ obtained from both CFDShip-Iowa V5.5 and STAR-CCM+. As shown in the figure, the results of both codes show similar trends, although noticeable differences can also be observed. Both the thrust and torque coefficients computed using CFDShip-Iowa V5.5 are lower compared to STAR-CCM+, which is consistent with the non-cavitating cases. The propeller open-water efficiencies computed using both solvers are close.

In general, the thrust and torque coefficients and the propeller efficiency increase and then decrease with the advance ratio. This trend differs from the fully wetted case without cavitation effects, where the thrust and torque coefficients decrease, and the propeller efficiency increases with advance ratios in the range of $J = 0.8$ to 1.4. At the low advance ratio $J = 0.8$, the blade section experiences a high angle of attack, resulting in a significant pressure drop on the suction side of the blade. This leads to the development of a large cavitation region, which consequently reduces both the thrust and torque. As the advance ratio increases, both the angle of attack and the pressure drop decrease, which results in a reduction in cavitation size and intensity. Consequently, the influence of cavitation on thrust and torque diminishes, and both the thrust and torque decrease from $J = 1.1$ to 1.4, similar to the non-cavitating cases. Compared to the non-cavitating conditions, both thrust and torque coefficients significantly decrease, especially at $J = 0.8$. As shown in Tables 2–4, the averaged reductions in thrust, torque, and propeller efficiency are 30.2%, 25.0%, and 7.6%, respectively, for the CFDShip-Iowa simulations. For STAR-CCM+, the corresponding averaged reductions are 23.4%, 14.5%, and 10.4%, respectively. Note that the reduction and change in the three variables at $J = 1.1$ are much smaller compared to the other two J values. This is because the $J = 1.1$ is close to the design value.

Figure 8 presents the results at the low cavitation number of 0.274. The simulation results from CFDShip-Iowa V5.5 generally agree with those from STAR-CCM+, except for the propeller efficiency (η), where CFDShip-Iowa V5.5 shows a much lower value at $J = 1.4$. The predicted trends of the thrust, torque, and propeller efficiency using CFDShip-Iowa V5.5 differ from those obtained by STAR-CCM+. For CFDShip-Iowa V5.5, all the variables increase first and then decrease with the advance ratios, which is similar to the results of the large cavitation number of $\sigma = 1.095$. As for STAR-CCM+, the thrust and torque coefficients decrease, and the propeller efficiency generally increases with advance ratios, which is similar to the non-cavitating cases. These differences are probably due to the blade-to-blade interactions, which are not captured in STAR-CCM+.

For the CFDShip-Iowa simulations (see Tables 2–4), the thrust, torque, and propeller efficiency are reduced on average by 63.4%, 49.0%, and 30.1%, respectively. For STAR-CCM+, the averaged reductions are 60.5%,52.8%, and 16.1% for the thrust, torque, and efficiency, respectively. The averaged reductions of both codes are close, except for the propeller efficiency, as also shown in Figure 8. Note that the changes in the three variables at $J = 1.1$, which is close to the design value, are smaller compared to those at $J = 0.8$ and $J = 1.4$.

3.2. Cavitation Profiles and Pressure Distributions

Figure 9 shows the cavitation vapor profiles for $\sigma = 1.095$ computed using both CFDShip-Iowa V5.5 and STAR-CCM+. As shown in the figures, the overall cavitation patterns are similar for both codes, where the cavitation size decreases with the advance ratios. CFDShip-Iowa V5.5 shows relatively more violent patterns compared to those of STAR-CCM+. At $J = 0.8$ and 1.1, cavitation shedding towards the pressure side can be observed in the CFDShip-Iowa V5.5 simulations, and the cavitation is mainly on the suction side of the blades in the STAR-CCM+ simulations.

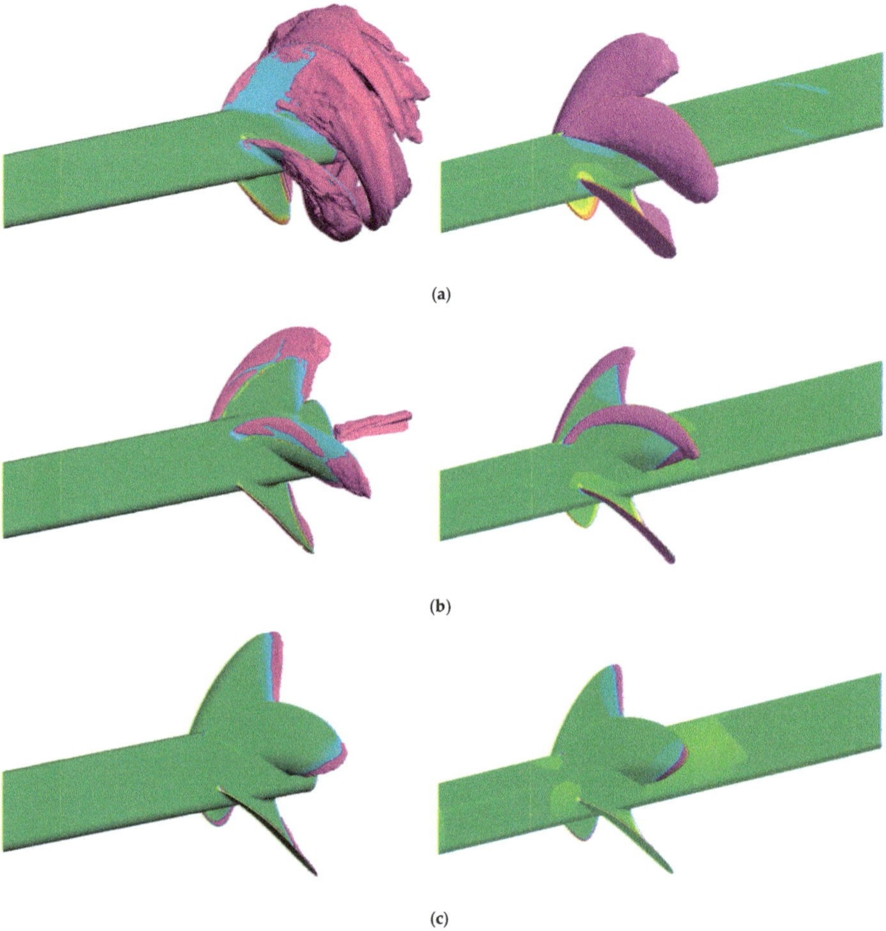

Figure 9. Vapor-phase iso-surface (vapor volume fraction of 0.1) for $\sigma = 1.095$. (**a**) $J = 0.8$; (**b**) $J = 1.1$; (**c**) $J = 1.4$. Left: CFDShip-Iowa V5.5; right: STAR-CCM+, rotated and duplicated from the single-blade results.

Pressure distributions on the surface of the propeller blade for $\sigma = 1.095$ are shown in Figures 10 and 11 for the suction and pressure sides, respectively. Generally, the results of both CFDShip-Iowa V5.5 and STAR-CCM+ show similar trends with advance ratios. The pressure on the suction side of the blade increases with J values, which is consistent with the cavitation size shown in Figure 9. For the pressure side, the pressure decreases with the advance ratios. Moreover, on the pressure side, STAR-CCM+ shows larger regions of high pressure than CFDShip-Iowa V5.5, especially at $J = 0.8$ and 1.1. This is due to the blade-to-blade cavitation interactions, where the cavity generated on the neighboring blade is extended onto the pressure side of the blade. As mentioned in the previous section, a single propeller blade with periodic boundary conditions is used for the STAR-CCM+ simulations, where the blade-to-blade interactions are not considered.

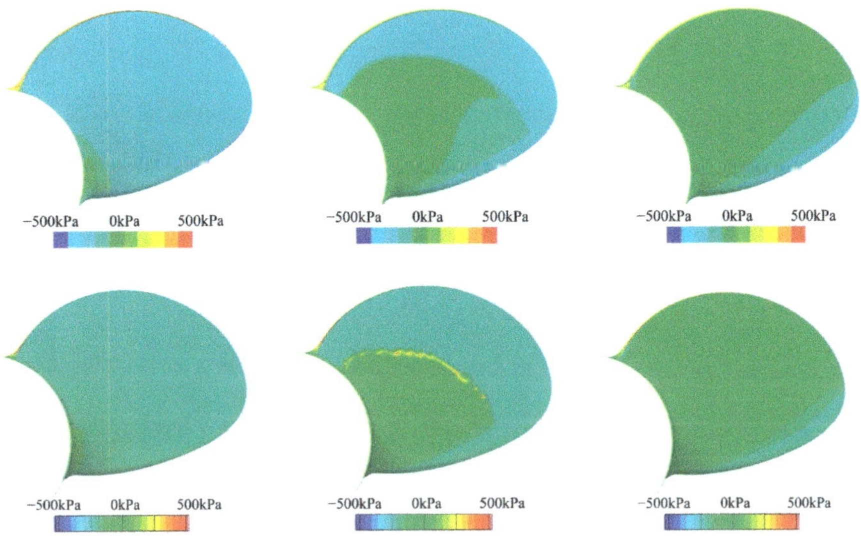

(a) (b) (c)

Figure 10. Pressure distributions on the suction side of the propeller blade for $\sigma = 1.095$. (**a**) $J = 0.8$; (**b**) $J = 1.1$; (**c**) $J = 1.4$. Top: STAR-CCM+; bottom: CFDShip-Iowa V5.5.

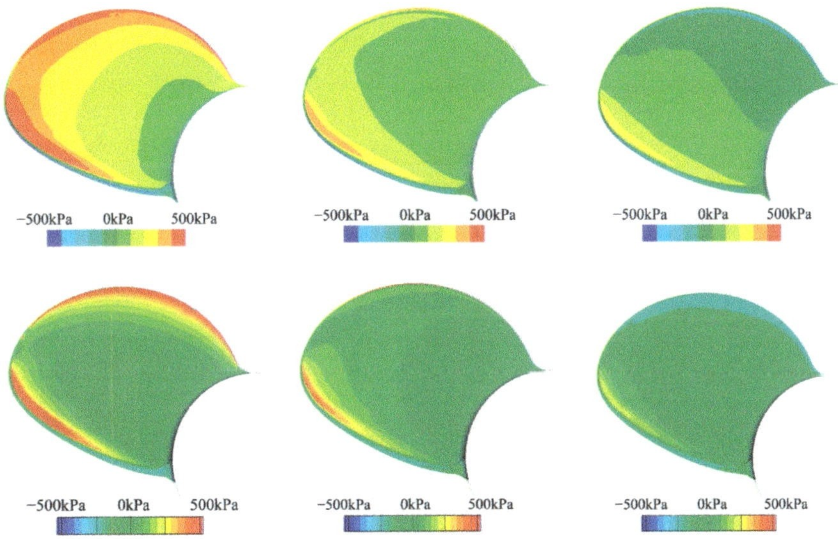

(a) (b) (c)

Figure 11. Pressure distributions on the pressure side of the propeller blade for $\sigma = 1.095$. (**a**) $J = 0.8$; (**b**) $J = 1.1$; (**c**) $J = 1.4$. Top: STAR-CCM+; bottom: CFDShip-Iowa V5.5.

Figure 12 shows the cavitation vapor profiles for $\sigma = 0.274$. The cavitation is more violent with a much larger size than that computed at the large cavitation number $\sigma = 1.095$.

The cavitation size also decreases with advance ratios, which is clearly shown in the simulation results of STAR-CCM+. For CFDShip-Iowa V5.5, the cavitation wraps around the entire propeller and is swept downstream for all the J values, which is observed only at $J = 0.8$ in the STAR-CCM+ simulations. Figure 13 shows the underwater high-speed video images of a propeller mounted on a production outboard motor, with a design similar to the one used in this study. The tests were conducted at speeds of 60 mph (26.8 m/s) and 70 mph (31.3 m/s), corresponding to cavitation numbers of $\sigma = 0.275$ and $\sigma = 0.202$, respectively. The propeller is partially submerged, and exhaust gas emanating from the back of the gearcase housing through the gap between housing and propeller is also present. At high speeds, this exhaust forms a sheath along the hub. As a result, the blades experience a mixture of cavitation, air, and exhaust, rather than pure cavitation, as in the simulations. Nevertheless, the images appear to support the presence of relatively large cavitation volumes, similar to those predicted by the CFDShip-Iowa simulations and significantly greater than those from the STAR-CCM+ simulations.

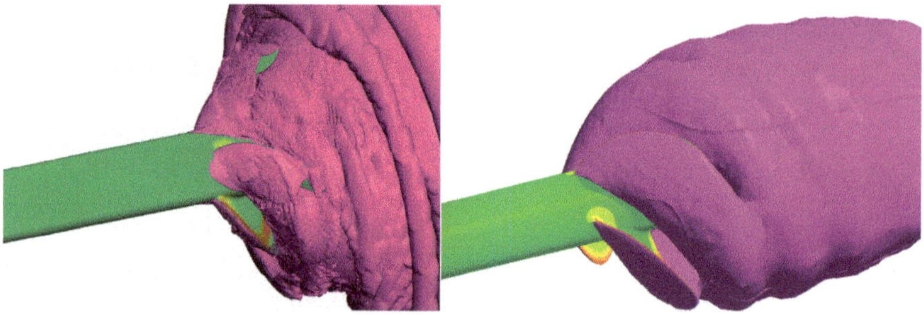

(a)

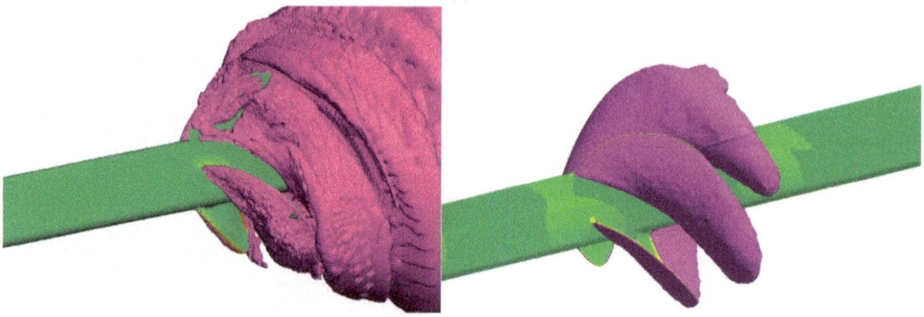

(b)

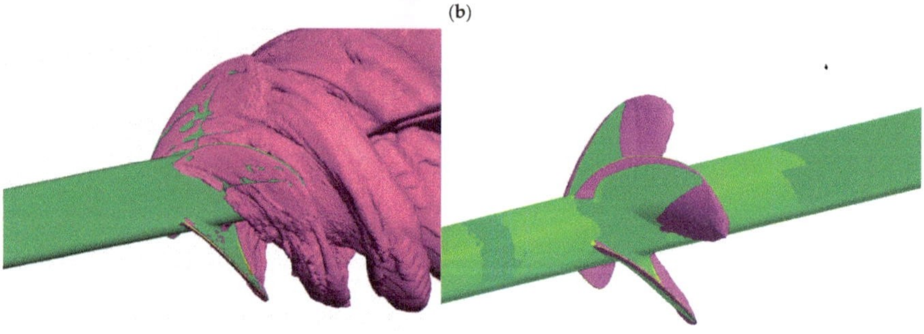

(c)

Figure 12. Vapor-phase iso-surface (vapor volume fraction of 0.1) for $\sigma = 0.274$. (**a**) $J = 0.8$; (**b**) $J = 1.1$; and (**a**,**c**) $J = 1.4$. Left: CFDShip-Iowa V5.5; right: STAR-CCM+, rotated and duplicated from the single-blade results.

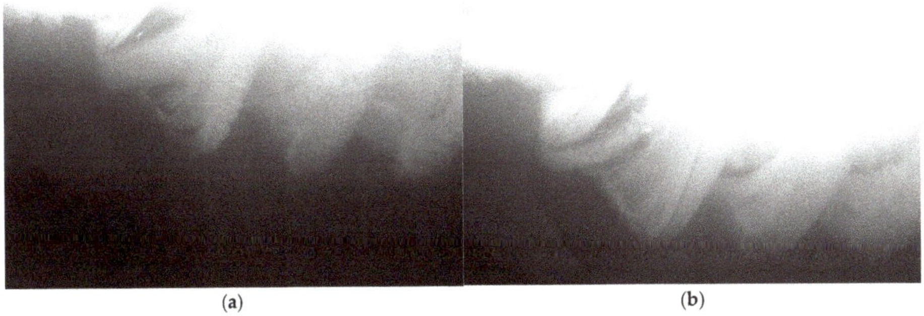

Figure 13. Experimental images from high-speed digital video. (**a**) $\sigma = 0.275$; (**b**) $\sigma = 0.202$.

The pressure distribution on the suction and pressure sides of the blade for $\sigma = 0.274$ is shown in Figures 14 and 15, respectively. Except for STAR-CCM+ at $J = 1.4$, where a small region of high pressure is observed on the suction side, the pressure is almost constant for all the cases, which indicates that entire blade is cavitated. This is consistent with the cavitation profiles shown in Figure 12. For the pressure side, the pressure decreases with advance ratios. STAR-CCM+ shows larger high-pressure regions than CFDShip-Iowa V5.5. A clear low-pressure distribution at the tip of the blade can be observed at $J = 0.8$ in the simulation results of CFDShip-Iowa V5.5. Again, this is due to the blade-to-blade cavitation interactions that are not captured in the STAR-CCM+ simulations.

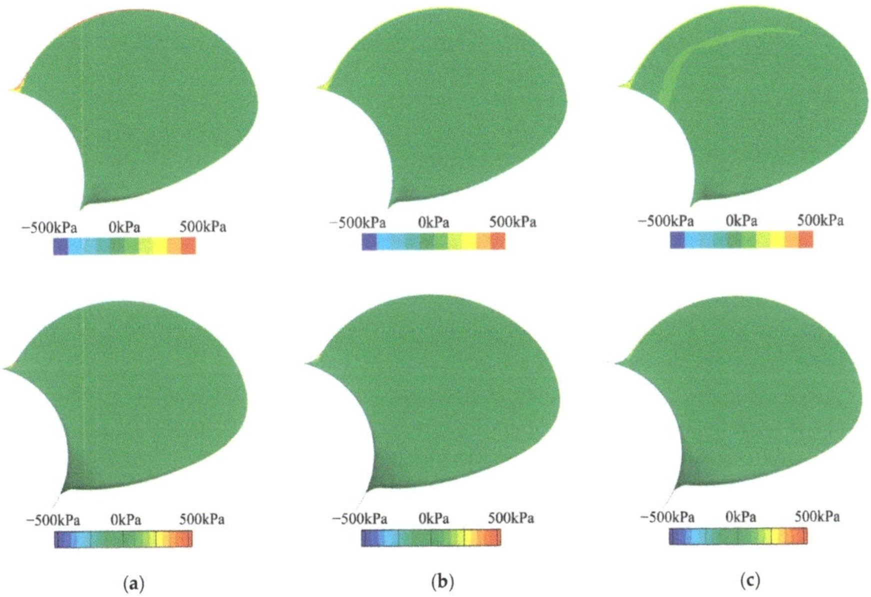

Figure 14. Pressure distributions on the suction side of the propeller blade for $\sigma = 0.274$. (**a**) $J = 0.8$; (**b**) $J = 1.1$; (**c**) $J = 1.4$. Top: STAR-CCM+; bottom: CFDShip-Iowa V5.5.

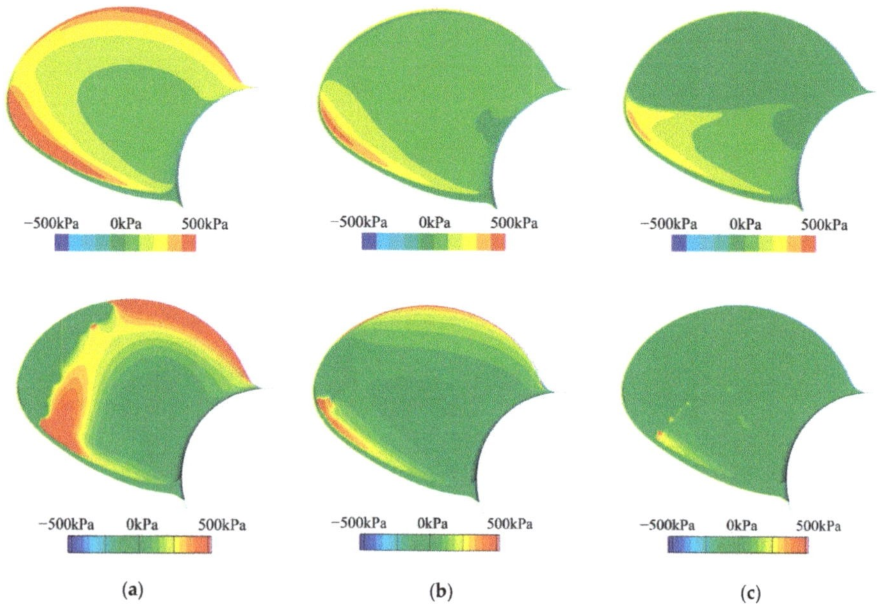

Figure 15. Pressure distributions on the pressure side of the propeller blade for $\sigma = 0.274$. (**a**) $J = 0.8$; (**b**) $J = 1.1$; (**c**) $J = 1.4$. Top: STAR-CCM+; bottom: CFDShip-Iowa V5.5.

4. Conclusions

Numerical simulations of an open-water propeller with cavitation effects are performed using CFDShip-Iowa V5.5 for a high-speed planing hull at a high Froude number of 1.84. The computed results are compared with those obtained from the simulations using STAR-CCM+. The overall trends of the computed thrust, torque, propeller efficiency, cavitation profile, and pressure distribution from both codes are consistent, except for the low cavitation number $\sigma = 0.274$, where the propeller efficiency is much lower than that obtained with STAR-CCM+.

For the fully wetted conditions without cavitation effects, the thrust and torque coefficients decrease, while the propeller efficiency increases with the advance ratio. Compared to the non-cavitating conditions, both thrust and torque coefficients significantly decrease with cavitation effects. For the large cavitation number $\sigma = 1.095$, the thrust, torque, and propeller efficiency increase and then decrease with the advance ratio. For the small cavitation number $\sigma = 0.274$, all the variables increase first and then decrease with the advance ratios for CFDShip-Iowa, and the thrust and torque coefficients decrease, and the propeller efficiency generally increases with advance ratios for STAR-CCM+. The cavitation is more violent in CFDShip-Iowa than STAR-CCM+ at $\sigma = 0.274$ due to the blade-to-blade interactions, which are not captured in the STAR-CCM+ simulations since a single blade is used.

Additional factors, including geometry and grids, mathematical models, and numerical methods, can also influence the simulation results and account for the observed differences between simulations of the two codes. Further investigations will be considered in future work. Self-propulsion and acceleration simulations of the GPPH using the actual propeller and gear case geometries will be conducted.

Author Contributions: Conceptualization, F.S. and J.S.; methodology, S.P., Z.W. and A.G.; software, Z.W.; validation, S.P., Z.W. and A.G.; formal analysis, S.P., Z.W. and J.S.; investigation, S.P. and A.G.;

resources, F.S.; data curation, S.P.; writing—original draft preparation, S.P. and Z.W.; writing—review and editing, Z.W.; visualization, S.P., A.G. and J.S.; supervision, F.S.; project administration, F.S.; funding acquisition, F.S. All authors have read and agreed to the published version of the manuscript.

Funding: This research was funded by the US Office of Naval Research, grant numbers N00014-20-1-2259 and N00014-22-1-2413.

Data Availability Statement: All the data are available in the paper.

Acknowledgments: This work is supported by the US Office of Naval Research grants N00014-20-1-2259 and N00014-22-1-2413 under the administration of Robert Brizzolara. The simulations presented in this work were performed at the Department of Defense (DoD) Supercomputing Resource Centers (DSRCs) through the High-Performance Computing Modernization Program (HPCMP).

Conflicts of Interest: Authors Andrew Gunderson and John Scherer are employed by the company Mercury Marine. The remaining authors declare that the research was conducted in the absence of any commercial or financial relationships that could be construed as a potential conflict of interest.

References

1. Olofsson, N. Force and Flow Characteristics of a Partially Submerged Propeller. Ph.D. Thesis, Chalmers University of Technology, Göteborg, Sweden, 1996.
2. Young, Y.L.; Kinnas, S.A. Fluid and Structural Modeling of Cavitating Propeller Flows. In Proceedings of the 5th International Symposium on Cavitation, Osaka, Japan, 1–4 November 2003.
3. Nouroozi, H.; Zeraatgar, H. A reliable simulation for hydrodynamic performance prediction of surface-piercing propellers using URANS method. *Appl. Ocean Res.* **2019**, *92*, 101939. [CrossRef]
4. Yari, E.; Moghadam, A.B. BEM applied to the cup effect on the partially submerged propeller performance prediction and ventilation pattern. *J. Mar. Eng. Technol.* **2020**, *21*, 159–177. [CrossRef]
5. Yilmaz, N.; Atlar, M.; Khorasanchi, M. An improved Mesh Adaption and Refinement approach to Cavitation Simulation (MARCS) of propellers. *Ocean Eng.* **2019**, *171*, 139–150. [CrossRef]
6. Zhang, X.; Xu, Q.; Zhang, M.; Xie, Z. Numerical Prediction of Cavitation Fatigue Life and Hydrodynamic Performance of Marine Propellers. *J. Mar. Sci. Eng.* **2023**, *12*, 74. [CrossRef]
7. Vázquez-Santos, A.; Camacho-Zamora, M.; Hernández-Hernández, J.; Herrera-May, A.L.; Santos-Cortes, L.D.C.; Tejeda-del-Cueto, M.E. Numerical Analysis and Validation of an Optimized B-Series Marine Propeller Based on NSGA-II Constrained by Cavitation. *J. Mar. Sci. Eng.* **2024**, *12*, 105. [CrossRef]
8. Zhang, B.; Zheng, C.S.; Shao, X.M.; Deng, J. Multiscale modeling of propeller cavitation flows in open water. *J. Hydrodyn.* **2025**, *37*, 1–13. [CrossRef]
9. Ristea, M.; Popa, A.; Volintiru, O.-N. CFD Design Performance Analysis for a High-Speed Propeller. *Appl. Sci.* **2025**, *15*, 8754. [CrossRef]
10. Ghoniem, A.M.; Elazm, A.S.A.; Benaya, M.I.; Mohamed, S.A. Effect of salinity on marine propeller performance. *J. Phys. Conf. Ser.* **2023**, *2616*, 012016. [CrossRef]
11. Irannezhad, M.; Kjellberg, M.; Bensow, R.E.; Eslamdoost, A. Experimental and numerical investigations of propeller open water characteristics in calm water and regular head waves. *Ocean Eng.* **2024**, *302*, 117703. [CrossRef]
12. Oloan, A.F.N.; Ariana, I.M.; Baidowi, A. Open Water and Performance Analysis of Marine Propeller with PBCF Based CFD Method. In Proceedings of the 6th International Conference on Marine Technology, Surabaya, Indonesia, 27 November 2021.
13. Grlj, C.G.; Degiuli, N.; Farkas, A.; Martić, I. Numerical Study of Scale Effects on Open Water Propeller Performance. *J. Mar. Sci. Eng.* **2022**, *10*, 1132. [CrossRef]
14. Ali, M.A.; Peng, H.; Qiu, W. Propeller-hull interaction simulation for self-propulsion with sinkage and trim. *Phys. Fluids* **2024**, *36*, 025159. [CrossRef]
15. Stern, F.; Wang, Z.; Park, S.; Milano, C.; Gunderson, A.; Scherer, J.; Yasukawa, H.; Diez, M. Multi-Fidelity Full Throttle Acceleration from Rest for High-Speed Planing-Hulls with Engine Curves. In Proceedings of the 36th Symposium on Naval Hydrodynamics, Busan, Republic of Korea, 14–19 June 2026.
16. Wang, Z.; Stern, F. Volume-of-fluid based two-phase flow methods on structured multiblock and overset grids. *Int. J. Numer. Methods Fluids* **2022**, *94*, 557–582. [CrossRef]
17. Wang, Z.; Park, S.; Stern, F. Cavitation Implementation Algorithms Based on Pressure Projection Method for Incompressible Flows with Three-Phase Interactions. *Int. J. Numer. Methods Fluids* **2025**, *97*, 1057–1072. [CrossRef]

18. Zwart, P.J.; Gerber, A.G.; Belamri, T. A Two-Phase Flow Model for Predicting Cavitation Dynamics. In Proceedings of the International Conference on Multiphase Flow, Yokohama, Japan, 30 May–3 June 2004.
19. Menter, F.R. Two-equation eddy-viscosity turbulence models for engineering applications. *AIAA J.* **1994**, *32*, 1598–1605. [CrossRef]
20. Noack, R. SUGGAR: A general capability for moving body overset grid assembly. In Proceedings of the 17th AIAA Computational Fluid Dynamics Conference, Toronto, ON, Canada, 6–9 June 2005; p. 5117.
21. Schnerr, G.H.; Sauer, J. Physical and Numerical Modeling of Unsteady Cavitation Dynamics. In Proceedings of the 4th International Conference on Multiphase Flow, New Orleans, LA, USA, 27 May–1 June 2001.

Journal of
Marine Science and Engineering

MDPI

Article

Data-Driven Optimization of Voith-Schneider Tug Operations: Towards a Digital Twin Framework for Port Energy Management

Feliciano Fraguela , Fernando Mendizábal, José M. Pérez-Canosa and José A. Orosa *

Campus Industrial del Ferrol, CITENI, Escola Técnica Superior de Náutica e Máquinas, Universidade da Coruña (UDC), Paseo de Ronda, 51, 15011 A Coruña, Spain; feliciano.fraguela@udc.es (F.F.); f.mendizabal@udc.es (F.M.); jose.pcanosa@udc.es (J.M.P.-C.)
* Correspondence: jose.antonio.orosa@udc.es

Abstract: This study presents a data-driven methodology to optimize the operational efficiency of a tugboat equipped with a Voith-Schneider Propeller (VSP) based on full-scale fuel consumption and vessel performance data. The objective is to identify optimal combinations of engine RPM and propeller pitch to reduce fuel consumption during low-demand phases without compromising maneuverability. Sea trials were conducted under controlled conditions using a dual flowmeter system and onboard speed measurements. The data enabled the construction of performance curves, efficiency ratios, and interpolated maps of fuel consumption. Optimal configurations were identified across defined speed ranges, and continuous efficiency zones were visualized through iso-consumption and contour plots. The results reveal a nonlinear relationship between propeller pitch, speed, and fuel demand, with maximum efficiency occurring at medium-to-high pitch values and speeds between 3 and 6 knots. This methodology provides a replicable tool for energy management in port operations and supports informed decisions during accompanying operations and standby periods. Efficiency differences over 300% between RPM–pitch settings were found, highlighting the operational impact of informed configuration choices. Moreover, the structured dataset and visual analysis framework lay the groundwork for future digital twin models aimed at enhancing operational efficiency in VSP-powered tugboats.

Keywords: fuel consumption; propeller pitch; Voith-Schneider propulsion; energy efficiency; tugboats; port operations; on-board data; operational modeling; decision support; digital twin

1. Introduction

Improving energy efficiency in maritime transport is a key objective in the global transition toward more sustainable logistics and reduced environmental impact. In recent years, international regulations such as the Energy Efficiency Design Index (EEDI) and operational measures like the Ship Energy Efficiency Management Plan (SEEMP) have driven growing attention toward propulsion optimization, fuel consumption reduction, and vessel performance monitoring [1–4]. In this context, harbor and escort tugs—despite their relatively small size—play a critical role in port operations and short-range navigation, often under highly dynamic and demanding conditions. Moreover, these vessels typically operate in densely populated coastal urban areas, where their environmental footprint—particularly in terms of pollutant emissions—adds to the existing load borne by port cities [5–7].

Among the various propulsion systems available, the Voith-Schneider propeller (VSP) stands out for its high maneuverability and immediate thrust vector control, making it especially suitable for tugs operating in restricted environments [8]. However, the unique kinematics of the VSP system also pose specific challenges in terms of energy efficiency, particularly regarding the optimal selection of propeller pitch and RPM across different speed regimes. Recent numerical studies have confirmed the sensitivity of VSP hydrodynamic performance to changes in operating parameters, reinforcing the need for full-scale data to guide optimization efforts [9,10].

The Voith-Schneider Propeller (VSP), technically classified as a cycloidal propulsion system, consists of a set of vertical blades mounted on a rotating disk. Unlike conventional propellers, each blade continuously adjusts its pitch as the disk rotates, allowing the generation of thrust in any horizontal direction without changing the orientation of the vessel. This mechanism provides full 360-degree thrust vectoring through mechanical control of blade angles.

While the hydrodynamic behavior and control strategies of cycloidal propulsion systems have been extensively studied through simulations and towing tank experiments [9,10], the availability of empirical data from full-scale operational settings remains limited. Most previous research has focused on idealized or controlled environments, often neglecting the complexity and variability of real navigational conditions. This empirical gap restricts the practical applicability of theoretical findings, especially when seeking to optimize energy performance under actual service demands. Recent studies have highlighted the importance of incorporating full-scale real-world measurements to improve the reliability of performance assessments and support operational decision-making in complex marine systems [3,11]. Additionally, recent research has explored the use of numerical simulations to optimize the efficiency of Voith-Schneider propulsion systems [12], reinforcing the relevance of nonlinear hydrodynamic behavior and control sensitivity.

Although digital twin applications in the maritime sector have grown significantly in recent years, most published studies focus on general-purpose ships, maintenance optimization, or hull performance modeling [13–15]. Some recent developments have addressed full-electric harbor crafts using integrated systems models [16], but empirical or digital twin-based studies specific to Voith-Schneider propulsion systems in tugboats remain scarce. For instance, while experimental hydrodynamic analyses of ASD tugs exist [17], equivalent real-time or data-driven frameworks for VSP-equipped vessels are lacking. These simulation-based approaches underscore the value of full-scale empirical data as a foundation for future digital twin frameworks aimed at supporting energy-aware operational strategies in tugboat applications [18,19].

Given the scarcity of digital twin studies specifically targeting VSP-equipped tugboats, this work offers a novel contribution by applying full-scale empirical data to support the development of operational efficiency models suitable for integration into future decision-support systems for harbor towing.

This study addresses this gap by presenting a methodology based on full-scale trials aboard a Voith-Schneider tugboat. Using synchronized measurements of vessel speed, fuel consumption, propeller pitch, and RPM, we aim to identify the operational configurations that maximize energy efficiency across the full range of service conditions. The main contributions include (i) the development of interpolated fuel consumption and efficiency maps, (ii) the definition of optimal propulsion regimes segmented by speed range, and (iii) the provision of practical guidelines for improving energy management in port operations. By relying on empirical data and applying visual analytics, this approach provides a replicable framework for enhancing the operational efficiency of vessels equipped with VSP systems.

2. Materials and Methods

This study was conducted using a Voith Water Tractor-type tugboat equipped with a cycloidal propulsion system (VSP) (Voith Turbo Schneider Propulsion GmbH & Co. KG, Heidenheim an der Brenz, Germany). The name of the vessel is not disclosed for operational confidentiality reasons; however, the relevant technical specifications are provided. The tug has an overall length of 29.5 m, a beam of 8.5 m, and a gross tonnage of approximately 300 GT. It is powered by two Caterpillar 3508 engines delivering a total power of 1600 kW, coupled to a dual cycloidal propulsion system. The VSP system allows for independent control of both the propeller's revolutions per minute (rpm) and the propeller pitch, enabling a detailed analysis of energy efficiency under different propulsion configurations.

2.1. Test Area and Environmental Conditions

The measurements were conducted in the estuary of Ferrol, located in the northwest of Spain, an estuarine environment selected for its natural protection from ocean swell and relatively stable tidal conditions. This location was chosen to minimize the influence of environmental disturbances, effectively functioning as an open-air experimental basin.

Figure 1 shows the track followed by the tug during the trials, where round-trip runs were carried out along the same axis for each propulsion setting. Figure 2 illustrates the typical behavior of tidal currents in the area, which was used as a reference to select time windows with minimal hydrodynamic interference.

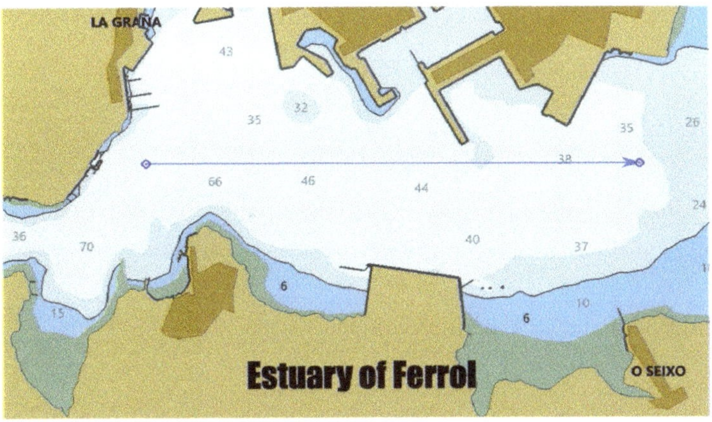

Figure 1. Track of the test runs performed by the tug in the estuary of Ferrol, showing the repeated route used for each measurement.

The four tide charts presented in Figure 2 correspond to average tidal cycles in the port of Ferrol. Spring tides—typically occurring near the new and full moons—were deliberately avoided due to their stronger currents and greater hydrodynamic impact on the vessel's movement.

All tests were conducted during slack tide periods to minimize tidal current interference. These slack windows generally occur near high and low tide. In this study, data were collected specifically around low tide slack periods, which exhibited near-zero current velocities and unstable flow directions, providing the most hydrodynamically neutral conditions for performance measurements. All trials were also carried out under light wind conditions.

For each operating point—defined by a specific combination of RPM and propeller pitch—two runs were conducted: one outbound and one return leg, in opposite directions relative to the wind and current. This procedure allowed for external influences to be compensated through averaging. Moreover, no active directional control was applied during navigation. The vessel proceeded in a straight line under free drift, avoiding any thrust vectoring inputs that could have introduced additional resistance or influenced fuel consumption.

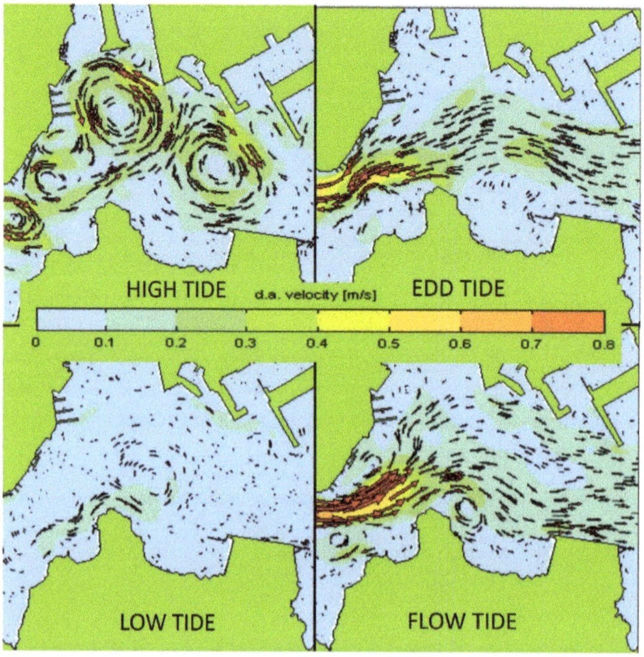

Figure 2. Tidal current behavior in the estuary of Ferrol, used to select optimal testing windows with minimal environmental interference.

2.2. Instrumentation and Data Acquisition

Fuel consumption was measured using two positive displacement flowmeters (J5015PT2, DN15, maximum capacity 50.0 L/min, ±0.2% accuracy) (Oval Corporation, Tokyo, Japan), installed on the inlet and return fuel lines of each engine. The selection and installation of these flowmeters required careful consideration of flow range, available space, and compatibility with marine diesel fuel. During early testing, one of the sensors experienced temporary clogging due to fuel contamination, emphasizing the need for pre-filtration and regular maintenance access. These operational issues were mitigated by installing upstream protective filters and ensuring physical accessibility for inspection. Effective fuel consumption was measured using two displacement flowmeters installed in the fuel supply and return lines. The FCM2 monitoring unit recorded both flow rates and automatically computed the net fuel consumption by calculating the difference between them. It also logged additional parameters such as fuel temperature in real time. The vessel's average speed during each test run was measured using its on-board GPS-based navigation system, while the values of propeller pitch and RPM were recorded from the bridge during each trial. Figure 3 provides a simplified schematic of the on-board data acquisition system used during the tests.

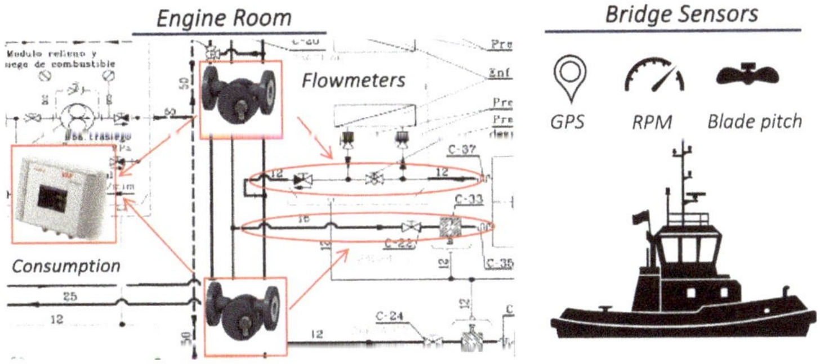

Figure 3. Schematic diagram of the on-board data acquisition system.

2.3. Measurement Protocol

Nine RPM settings were defined, ranging from 700 to 1500 rpm in 100 rpm intervals, which corresponds to the operational range of the Caterpillar 3508 engine installed on the tugboat, along with ten distinct propeller pitch positions, resulting in a complete matrix of operating combinations. For each test configuration—defined as a unique combination of RPM and propeller pitch—three full round-trip runs were performed within the designated test area.

To ensure repeatability and minimize the influence of transient variations, six independent measurements were obtained for each configuration (three outbound and three return runs), and the average values were used to define the representative result for each measured variable. This procedure helped reduce the influence of external factors such as wind, residual currents, and engine load fluctuations, thereby enhancing the reliability of the data used in subsequent analysis. The experimental strategy aligns with established recommendations for full-scale performance testing under real operating conditions [20]. For each run, data recording for fuel consumption began approximately one minute after the vessel had reached a steady-state speed in order to avoid transient effects associated with acceleration and stabilization phases.

Although the vessel is equipped with two identical main engines operating symmetrically, fuel consumption was monitored only on the port-side engine. Since the load was evenly distributed and both engines operated under identical control settings during straight-line navigation, the measurements were considered representative of the overall propulsion system behavior. All efficiency calculations and consumption maps were based on this single-engine data, under the assumption of symmetric performance.

2.4. Data Processing and Analysis

Data processing was performed using Python 3.12.4. Propulsive efficiency was calculated as the ratio between the vessel's average speed (in knots) and fuel consumption (in L/min), following a simplified metric commonly used in operational performance studies [21]. This operational ratio, while not equivalent to conventional efficiency indicators based on delivered thrust or shaft power, offers a practical method for assessing relative performance under real navigation conditions.

In Voith-Schneider propulsion systems, the installation of torque sensors presents significant technical challenges due to the enclosed hydromechanical configuration, which prevents direct access to a traditional transmission shaft. Moreover, thrust measurements can only be obtained through bollard pull tests conducted in static and controlled environments, which do not reflect the vessel's behavior under dynamic sailing conditions.

The main advantage of this simplified metric lies in its ability to consistently compare different propulsion configurations using data readily available on board, especially in vessels without advanced instrumentation. However, its main limitation is that it does not account for internal mechanical losses or actual thrust output and therefore should not be interpreted as an absolute measure of propulsive efficiency. It should be understood as a relative indicator, primarily suited for internal optimization within the same vessel, although its comparative value may extend to similar tugboats equipped with VSP systems and comparable dimensional and operational profiles.

To visualize the relationships between variables and capture the overall behavior of the propulsion system, interpolated surfaces were generated using cubic interpolation implemented through the griddata() function from the SciPy library [22]. These interpolations were performed over a regular 100 × 100 mesh constructed using the observed ranges of vessel speed and propeller pitch.

To validate the consistency of the experimental data and justify the use of average values in the analysis, the repeatability of each test point was assessed using statistical dispersion metrics. For each combination of RPM and propeller pitch, the standard deviation (STDEV) and the standard error of the mean (SEM) were calculated based on the six replicated measurements (three outbound and three return runs). The mean STDEV across all test configurations was 0.537 knots for vessel speed and 0.0187 kg/h for fuel consumption, while the corresponding SEM values were 0.219 knots and 0.0078 kg/h, respectively. These levels of dispersion are considered acceptable for full-scale trials conducted under real navigation conditions and confirm that the average values derived from the six measurements reliably represent the vessel's performance under each tested configuration. This enabled the creation of contour plots, iso-consumption maps, and three-dimensional efficiency surfaces using standard visualization tools [23]. The resulting data were segmented by speed range to identify optimal operating zones, and the most efficient configurations within each range were extracted to support comparative evaluation and operational guidance.

3. Results

This section presents the results derived from the analysis of the data recorded during the sea trials. Graphical representations are used to characterize the influence of propeller pitch and RPM on vessel speed and fuel consumption. Subsequently, propulsive efficiency curves were calculated, and interpolated surfaces were generated to identify optimal operating zones. Finally, the most efficient combinations were selected for each speed range, allowing for a detailed classification of the propulsion system's energy performance.

3.1. Relationship Between Propeller Pitch and Vessel Speed

Figure 4 shows the evolution of vessel speed as a function of propeller pitch across different engine speeds. A clear nonlinear trend is observed: at each RPM level, increasing the propeller pitch results in a higher average speed, although with a progressively reduced gain at higher pitch values. This diminishing return effect is more evident at low rpm (e.g., 700–900), where speed tends to stabilize beyond 6–7. In contrast, at higher rpm values (1300–1500), the increase in speed remains more linear over the full pitch range. These results highlight the combined influence of propeller pitch and engine speed on propulsion behavior and support a detailed optimization strategy based on operational conditions. This nonlinear behavior, as well as the saturation trend observed at higher propeller pitches, is consistent with hydrodynamic principles described in the technical literature on marine propellers [21].

These findings support the need to identify pitch thresholds that optimize vessel performance without incurring unnecessary energy input, a critical consideration for energy-efficient tug operation.

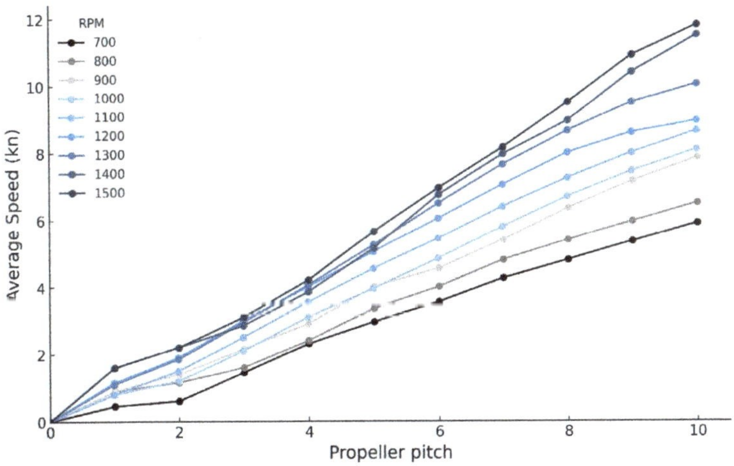

Figure 4. Relationship between propeller pitch and vessel speed across RPM levels.

3.2. Relationship Between Propeller Pitch and Fuel Consumption

Figure 5 presents the variation in average fuel consumption as a function of propeller pitch across the different RPM settings. In all cases, fuel consumption increases progressively with pitch, although the rate of increase varies depending on the engine speed. At low rpm (700–900), the rise in consumption is more moderate and tends to stabilize beyond a certain pitch threshold. In contrast, at higher rpm values (1300–1500), consumption increases more sharply and continuously across the entire pitch range, with a noticeable change in slope occurring beyond. This behavior suggests a growing propulsive load and a potential decrease in energy efficiency at high-pitch settings. These results are essential for identifying operating settings that minimize fuel usage and support energy-efficient performance. This increasing and nonlinear fuel consumption behavior, associated with both propeller pitch and RPM, is consistent with experimental observations of cycloidal propulsion systems under varying operating conditions [10].

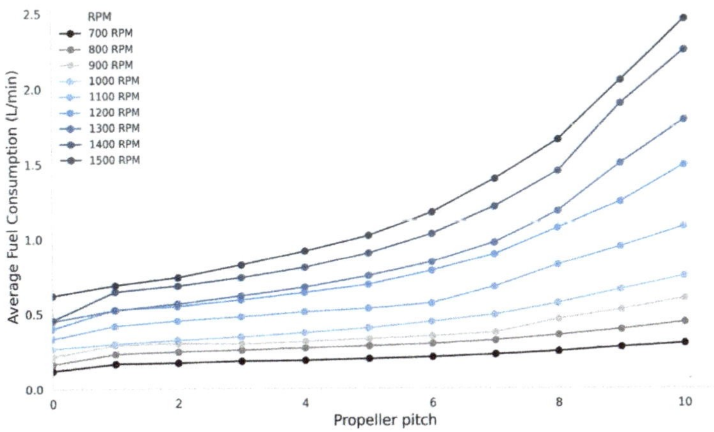

Figure 5. Fuel consumption (L/min) as a function of propeller pitch for different RPM settings.

Understanding these consumption trends is essential to prevent operational regimes where increased propeller pitch yields diminishing returns in speed but disproportionately higher fuel usage [11].

3.3. Propulsive Efficiency Curves as a Function of Vessel Speed

Figure 6 presents the propulsive efficiency, defined as the ratio between average speed and fuel consumption, as a function of vessel speed for each propeller pitch considered. Each curve represents a fixed propeller pitch, and each data point along the curve corresponds to a different engine RPM, ranging from 700 rpm (leftmost point) to 1500 rpm (rightmost point). In general, efficiency increases with speed up to a certain point, reaches a local maximum, and then tends to stabilize or slightly decrease.

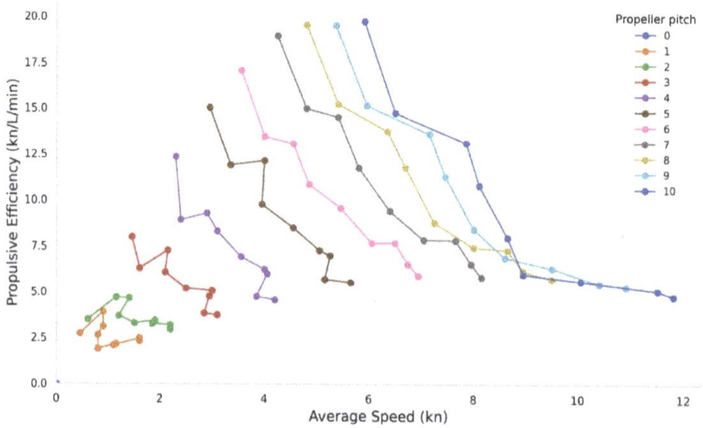

Figure 6. Propulsive efficiency (kn/(L/min)) as a function of vessel speed for different propeller pitches.

Intermediate propeller pitches (between 6 and 8) exhibit the highest efficiency values, particularly within the speed range of 3 to 6 knots, where the propulsion system achieves a favorable balance between speed and fuel consumption. In contrast, low propeller pitches (1 to 3) show reduced efficiency across all speeds due to their limited ability to generate effective propulsive response. At the other extreme, higher propeller pitches (9 and 10) display a progressive decline in efficiency at higher speeds, likely due to the substantial increase in fuel consumption required to maintain those regimes.

This representation enables the direct identification of optimal combinations of speed and propeller pitch and serves as a valuable tool for designing energy-efficient operational strategies.

Propulsive efficiency, defined as the ratio of speed to fuel consumption, offers a simple yet useful metric for evaluating the operational performance of the system in relative terms, in line with other methodological approaches aimed at improving energy performance under real-world sailing conditions [3,11,24].

3.4. Three-Dimensional Map of Fuel Consumption

Figure 7 presents a three-dimensional surface plot of fuel consumption as a function of the vessel's average speed and propeller pitch. This visualization clearly illustrates how both parameters interact to determine the energy demand of the propulsion system. Fuel consumption increases progressively with both speed and propeller pitch, with a more pronounced rise observed at higher values of these variables.

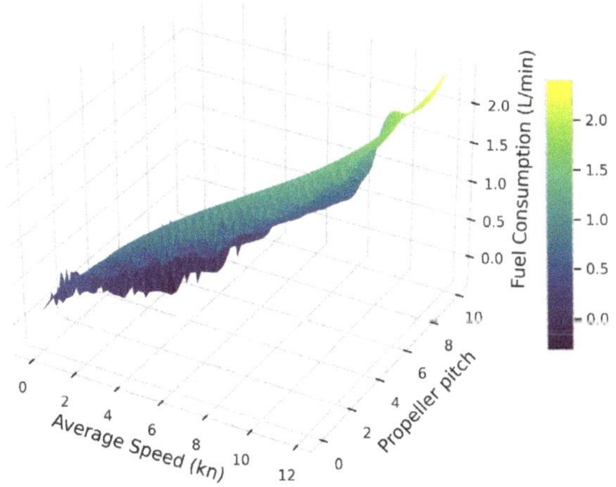

Figure 7. Interpolated 3D surface plot of fuel consumption (L/min) as a function of vessel speed and propeller pitch.

The interpolated surface reveals a region of moderate consumption associated with mid-range propeller pitches and speeds, whereas the highest consumption gradients are concentrated in the upper-right area of the surface—corresponding to propeller pitches above 8 and speeds greater than 6 knots. This trend suggests a less efficient operating regime under maximum propulsive demand, which could be avoided by selecting combinations of speed and propeller pitch that yield lower energy costs for comparable performance.

This graphical analysis provides a valuable tool for identifying suboptimal operating zones and supporting the design of navigation strategies focused on energy efficiency.

3.5. Iso-Consumption Curves

Figure 8 presents the iso-consumption curves, which represent lines of equal fuel consumption as a function of vessel speed and propeller pitch. This two-dimensional representation allows for a clear identification of operating combinations that result in the same energy demand, making it particularly useful for comparing different propulsion scenarios.

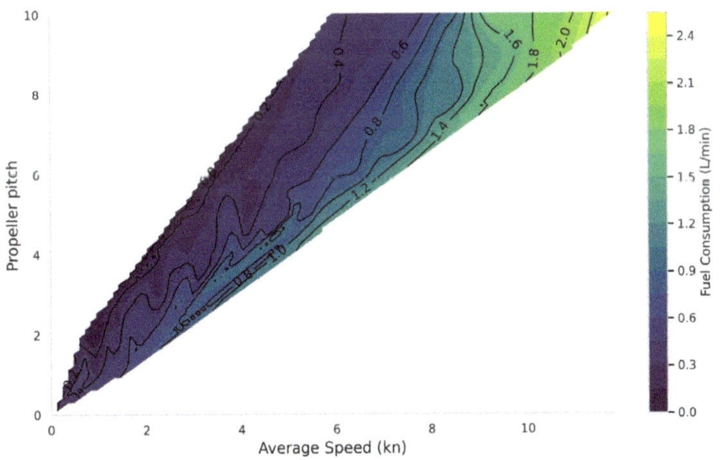

Figure 8. Iso-consumption curves derived from interpolated fuel consumption data as a function of vessel speed and propeller pitch.

A general upward trend is observed, with fuel consumption increasing progressively as both speed and pitch increase. Iso-consumption lines are more widely spaced in the low-consumption region—corresponding to intermediate propeller pitches and moderate speeds—indicating a zone of higher relative efficiency. In contrast, in the upper-right area of the graph (high pitch and high speed), the curves are more closely packed, reflecting a sharper rise in fuel demand.

This visualization complements the previously analyzed three-dimensional consumption map, providing a more precise perspective for identifying efficiency thresholds and supporting real-time operational decision-making.

3.6. Optimal Combinations by Speed Range

To identify the most energy-efficient propulsion configurations across the vessel's full operational profile, the top three combinations with the highest efficiency (speed/fuel consumption) were determined within each speed range. Table 1 summarizes these results, grouped into six intervals: Very Low (0–2 kn), Low (2–4 kn), Moderate (4–6 kn), Medium-High (6–8 kn), High (8–10 kn), and Very High (>10 kn).

Table 1. Top three operating combinations per speed range, sorted by propulsive efficiency.

Speed Range (kn)	RPM	Propeller Pitch	Fuel Consumption (L/min)	Efficiency kn/(L/min)
Very Low (0–2)	700	3	0.18	8.00
	800	3	0.25	6.30
	800	2	0.24	4.72
Low (>2–4)	700	6	0.21	17.11
	700	5	0.20	15.06
	800	6	0.30	13.49
Moderate (>4–6)	700	10	0.30	19.79
	700	8	0.25	19.57
	700	9	0.27	19.56
Medium-high (>6–8)	800	10	0.44	14.79
	900	8	0.46	13.78
	900	9	0.52	13.65
High (>8–10)	1000	10	0.75	10.82
	1100	10	1.08	8.00
	1300	8	1.18	7.32
Very High (>10)	1300	10	1.79	5.61
	1400	9	1.90	5.47
	1500	9	2.06	5.30

The highest efficiency values are concentrated in the Low, Moderate, and Medium-High ranges, indicating that these operational zones yield a relatively greater energy performance. In these intervals, the most efficient combinations involve intermediate propeller pitches and moderate RPM, achieving efficient speeds while maintaining controlled fuel consumption. As target speed increases, progressively higher propeller pitches are required to sustain acceptable efficiency levels. In the upper ranges (High and Very High), the most efficient configurations reflect a trade-off between elevated speed and fuel economy, operating within a regime of a relatively lower efficiency. This detailed classification allows for the adaptation of propulsion strategies to different mission profiles, maximizing the tug's energy performance under varying navigation conditions.

This analysis transforms the technical information obtained into structured operational recommendations by speed range, representing a further step beyond previous studies focused solely on hydrodynamic characterization of the propulsion system. Such an applied approach is essential for improving energy performance under real operating conditions [11].

3.7. Optimal Operating Zones

Figure 9 presents an interpolated efficiency map as a function of average vessel speed and propeller pitch, allowing for a visual identification of the most favorable operating zones from an energy standpoint. The color scale represents efficiency values (defined as average speed divided by fuel consumption), enabling the delimitation of parameter combinations that maximize propulsive performance.

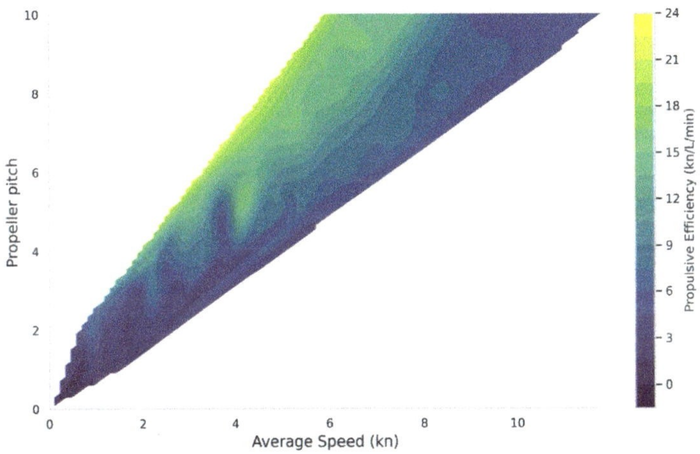

Figure 9. Interpolated propulsive efficiency map as a function of vessel speed and propeller pitch.

The map reveals a well-defined region of maximum efficiency centered around speeds between 3 and 6 knots and propeller pitches between 6 and 9. These operating conditions represent an optimal balance between vessel speed and fuel demand, forming the core of the efficient operating zone. Beyond these values—whether through increased speed or propeller pitch—efficiency progressively decreases, delineating areas of lower energy performance. Similarly, low speed and low pitch combinations also result in suboptimal use of the propulsion system.

This graphical representation synthesizes the preceding analyses and serves as a practical tool for quickly identifying energy-efficient operating configurations.

Unlike the discrete range-based analysis presented in the previous subsection, this representation enables the identification of continuous high-efficiency operating zones, supporting decision-making under varying navigation conditions.

4. Discussion

The sea trials and interpolated performance maps confirmed a nonlinear relationship between propeller pitch and vessel speed, with clear efficiency saturation at high pitch values and increasing fuel penalties at higher RPMs. While these findings are technically consistent with previous hydrodynamic studies on VSP systems [10], their significance in this study lies in their applicability to real-world tug operations.

From an operational standpoint, the identified patterns reveal efficiency thresholds that can guide low-consumption regimes during transit, standby, or escorting operations.

Recognizing these optimal pitch and speed combinations is critical for port operators aiming to reduce fuel costs without compromising the tug's responsiveness.

More importantly, the empirical maps developed here provide a visual basis that can be integrated into real-time decision support systems. These tools translate complex propulsion system behavior into clear and intuitive operating zones, facilitating decision-making both on board and at the fleet management level. As such, they not only support immediate improvements in energy efficiency but also form a functional building block for future digital twin systems, where real-time feedback and adaptive propulsion modeling can optimize port energy management.

The interpolated representations and iso-consumption curves reinforce these findings, enabling the identification of continuous high-efficiency zones beyond discrete data points. The consistency between graphical methods and efficiency metrics supports the robustness of the results and their practical application.

Unlike previous studies that focused on hydrodynamic characterization under controlled conditions, this work was developed using real data obtained during navigation, adding practical relevance and robustness. Moreover, the applied focus of the methodology allows it to go beyond technical analysis, offering operational criteria for optimizing the energy performance of the overall propulsion system.

The methodological approach—combining full-scale measurements with visual interpolations and speed range segmentation—enables concrete operational recommendations tailored to different mission profiles. This adaptability is particularly valuable in port contexts, where tugboats alternate between high-power maneuvers and prolonged low-demand phases.

Table 1 represents a novel contribution by translating full-scale experimental data into a clear and operational classification of optimal propulsion settings across different speed ranges. Unlike most previous studies focused on hydrodynamic simulations or idealized tests, this table offers a real-world decision-support tool by identifying the most energy-efficient combinations of engine RPM and propeller pitch for each navigation regime.

A particularly relevant example is the moderate speed range (4–6 knots), which corresponds to one of the most common operational phases for tugboats during transit or escorting. Within this interval, the most efficient configuration reached a value of 19.79 kn/(L/min), while the least efficient barely exceeded 4.5. This difference represents a gain of over 330% in efficiency between two possible settings within the same speed range. Access to this information is essential for selecting operational parameters that not only reduce fuel consumption significantly but also lower the vessel's environmental impact—highlighting the practical and scientific relevance of the findings presented in this study.

These results were obtained under carefully selected environmental conditions (slack tide, light winds, repeated course) to minimize external disturbances and isolate the vessel's intrinsic propulsion performance. Therefore, the configurations presented in Table 1 are valid under moderately stable operating conditions, where the influence of external forces is limited.

However, in more dynamic operating environments—such as those involving current, wind, or wave action—the vessel's speed for a given configuration may deviate from the reference scenario. In such cases, a new optimal configuration would need to be identified for the altered dynamic equilibrium. This underscores the value of the empirical maps as a baseline reference while also highlighting the importance of future research focused on real-time adaptation to varying external conditions.

Moreover, the structured dataset and interpolated efficiency maps developed in this study provide a robust foundation for the development of digital twin systems applied to tugboat operations. A digital twin in this context would require the integration of three essential components: (i) a real-time data acquisition module (including vessel speed, fuel flow rate, RPM, and propeller pitch), (ii) a validated behavioral model (such as the efficiency surfaces obtained in this work), and (iii) a bidirectional communication interface to enable live monitoring and operational decision support. In terms of data acquisition and integration, onboard sensors can be connected through standard marine protocols (e.g., CAN bus, NMEA 2000), with readings stored in a time-series database embedded in an industrial PC or PLC. The interpolated maps can be loaded as look-up tables and periodically updated with operational logs, enabling a semi-online model. More advanced implementations may include machine learning modules for continuous model refinement, as explored in recent real-time digital twin architectures [15]. These enhancements would enable dynamic simulation, predictive maintenance, and real-time optimization, aligning with the broader vision of intelligent and energy-efficient port tug operations [13,25].

Finally, the use of real-world data collected under navigation conditions, combined with interpolation techniques applied to key propulsion system variables, results in a methodological framework that is replicable for other vessels equipped with VSP systems. Although the results were obtained from a specific tugboat, the structural similarity shared by most of these vessels suggests that the identified patterns could be extrapolated—at least as reference guidelines—to other units with comparable characteristics. Given the minimal technical requirements to install flowmeters on existing vessels, the proposed approach is particularly suited for real-world adoption in port towing operations.

5. Limitations and Future Developments

Despite the robustness of the approach and the consistency of the results obtained, this study presents certain limitations worth highlighting. First, the trials were conducted in a specific navigation area selected for its low exposure to currents and wind, but not under fully controlled conditions. Although repeated measurements and averaging were employed to reduce variability, the real operational environment inevitably introduces some dispersion that may affect speed and fuel consumption readings. Additionally, although fuel consumption was monitored on a single engine, the symmetrical configuration and balanced operation of both main engines under steady-state conditions support the representativeness of the data. Nevertheless, monitoring both engines could further reduce uncertainty in future implementations.

Second, the analysis focused on straight-line navigation without rudder input to isolate the behavior of the propulsion system. This means that the conclusions are particularly valid for transit or accompanying operations phases and may not be directly applicable to operations involving frequent heading changes or intensive lateral thrust.

Lastly, while the results were derived from a single vessel, the methodology has proven to be replicable and adaptable. Future application of this approach to a fleet of tugs with similar VSP configurations would help validate the extrapolation of the identified patterns and support the development of more robust predictive models.

As for future work, the incorporation of environmental variables such as current, wind, and towed ship load is proposed to refine operational efficiency in more complex scenarios. Additionally, integrating this methodology into on-board decision support systems or fleet management platforms could enhance its practical applicability and contribute to reducing energy consumption in port operations. As a further step, the availability of extended datasets under a wider range of environmental conditions (e.g., varying wind and current scenarios) could enable the mathematical formulation of the optimization problem, paving the way for the use of formal optimization algorithms based on the empirical performance surfaces developed in this study.

In parallel, the applicability of this methodology to upcoming Voith electric propulsion systems (eVSP) could be investigated. The enhanced responsiveness and torque control of eVSP units may affect the efficiency behavior observed in conventional setups and offer new optimization possibilities.

Finally, although this study did not implement a digital twin system, the structured dataset and visual performance framework developed here represent a solid starting point for future digital twin models aimed at integrating operational data with simulation environments. Future work may explore the real-time synchronization of onboard measurements and predictive models to support advanced decision-making tools in VSP-powered tugboats.

6. Conclusions

- Full-scale trials allowed the identification of optimal combinations of vessel speed and propeller pitch in a tugboat with VSP propulsion, aiming to improve energy efficiency under real operating conditions. In typical transit speed ranges (e.g., 4–6 knots), the most efficient configuration outperformed the least efficient by over 300%, highlighting the operational impact of selecting appropriate settings.
- The results revealed a nonlinear relationship between propeller pitch, engine RPM, and fuel consumption. The highest efficiency was concentrated at medium-to-high pitch settings and medium cruising speeds, especially during low-demand phases such as standby and accompanying operations.
- The proposed methodology, based on empirical data and interpolation techniques, enables the generation of visual tools and performance indicators that are easily interpretable and useful for planning energy-efficient operating regimes.
- The approach is replicable and practical, supporting direct fuel savings in port operations without compromising the vessel's responsiveness during high-power maneuvers.
- Moreover, the performance surfaces and structured dataset developed in this study address a core component of future digital twin systems, paving the way for their implementation through real-time data integration and adaptive operational modeling.

Author Contributions: Conceptualization, J.A.O., F.F., J.M.P.-C. and F.M.; methodology, F.F., J.M.P.-C., J.A.O. and F.M.; software, F.M., J.M.P.-C., J.A.O. and F.F.; validation, J.A.O., J.M.P.-C., F.F. and F.M.; investigation, F.F., J.M.P.-C., J.A.O. and F.M.; resources, F.F., J.M.P.-C., F.M. and J.A.O.; data curation, F.F., J.A.O., J.M.P.-C. and F.M.; writing—original draft preparation, F.F., J.A.O., J.M.P.-C. and F.M.; writing—review and editing, F.F. and J.A.O.; supervision, F.F., J.M.P.-C. and J.A.O.; funding acquisition; F.M., F.F., J.M.P.-C. and J.A.O. All authors have read and agreed to the published version of the manuscript.

Funding: This research was funded by University of A Coruña grant number I000815.

Data Availability Statement: The data supporting the findings of this study are available from the corresponding author upon reasonable request. Some restrictions apply due to operational confidentiality agreements.

Acknowledgments: The authors would like to thank Voith Turbo GmbH for their technical support and knowledge sharing throughout the development of this study. This research was also supported by the company Mantenimientos y Servicios Tecman, S.L., through the collaboration agreement with Universidade da Coruña under contract INV02224, and by the Port Authority of Ferrol-San Cibrao.

Conflicts of Interest: The authors declare no conflicts of interest.

References

1. International Maritime Organization, Third IMO GHG Study 2014. 2014. Available online: https://www.imo.org/en/ourwork/environment/pages/greenhouse-gas-studies-2014.aspx (accessed on 22 July 2025).
2. International Maritime Organization, Resolution MEPC.203(62): Amendments to MARPOL Annex VI—Inclusion of Regulations on Energy Efficiency for Ships. 2011. Available online: https://wwwcdn.imo.org/localresources/en/KnowledgeCentre/IndexofIMOResolutions/MEPCDocuments/MEPC.203(62).pdf (accessed on 22 July 2025).
3. Bouman, E.A.; Lindstad, E.; Rialland, A.I.; Strømman, A.H. State-of-the-art technologies, measures, and potential for reducing GHG emissions from shipping—A review. *Transp. Res. Part D Transp. Environ.* **2017**, *52*, 408–421. [CrossRef]
4. Tadros, M.; Ventura, M.; Soares, C.G. Review of the IMO Initiatives for Ship Energy Efficiency and Their Implications. *J. Mar. Sci. Appl.* **2019**, *22*, 662–680. [CrossRef]
5. Saxe, H.; Larsen, T. Air pollution from ships in three Danish ports. *Environ. Sci. Policy* **2004**, *7*, 545–551. [CrossRef]
6. Jiang, L.; Kronbak, J.; Christensen, L. The costs and benefits of sulfur reduction measures: Sulfur scrubbers versus marine gas oil. *Transp. Res. Part D Environ.* **2014**, *28*, 19–27. [CrossRef]
7. Jaramillo, J.; Gutiérrez, J.; Macia, Y.M. Comparative Analysis of Carbon Intensity Indicators Applicable to Harbor Tugboats. *Sustainability* **2025**, *17*, 1706. [CrossRef]
8. Voith Turbo GmbH. *Voith Schneider Propeller—Maritime Applications and Performance*; Voith Turbo GmbH: Heidenheim, Germany, 2016.
9. Jürgens, D.; Palm, M.; Singer, S.; Urban, K. Numerical Optimization of the Voith-Schneider Propeller. *ZAMM–J. Appl. Math. Mech.* **2007**, *87*, 792–794. [CrossRef]
10. Bakhtiari, M.; Ghassemi, M. A 2.5D numerical study on open water hydrodynamic performance of a Voith Schneider propeller. *Mech. Ind.* **2019**, *20*, 617. [CrossRef]
11. Baldi, F.; Johnson, H.; Gabrielii, C.; Andersson, K. Energy and exergy analysis of ship energy systems—The case study of a chemical tanker. *Int. J. Thermodyn.* **2015**, *18*, 82–93. [CrossRef]
12. Liu, W.; Chen, Q.; Ma, C. Research on efficiency optimization of the voith-schneider propeller based on motion curve parameter control. *Ocean Eng.* **2024**, *299*, 117136. [CrossRef]
13. Madusanka, N.; Fan, Y.; Yang, S.; Xiang, X. Digital Twin in the Maritime Domain: A Review and Emerging Trends. *J. Mar. Sci. Eng.* **2023**, *11*, 1021. [CrossRef]
14. Assani, N.; Matić, P.; Katalinić, M. Ship's Digital Twin—A Review of Modelling Challenges and Applications. *Appl. Sci.* **2022**, *12*, 6039. [CrossRef]
15. Lee, J.-H.; Nam, Y.-S.; Kim, Y.; Liu, Y.; Lee, J.; Yang, H. Real-time digital twin for ship operation in waves. *Ocean Eng.* **2022**, *266*, 112867. [CrossRef]
16. Hadi, J.; Zhi Yung, T.; Kudupudi, R.; Taskar, B.; Tan, C.E.; Singh Chopra, P.; Peng, N.; Konovessis, D. Digital Twin Modelling for Full-Electric Harbour Craft: An Integrated Hydrodynamics and Systems Model. In *ASME 2024 43rd International Conference on Ocean, Offshore and Arctic Engineering, Proceedings of the Offshore Technology, Singapore, 9–14 June 2024*; ASME: New York, NY, USA, 2024; Volume 1, p. V001T01A046. [CrossRef]
17. Zheng, Y.; Yiew, L.; Jin, Y.; Ravindra, K.; Magee, A.R. Experimental and numerical study on seakeeping performance of a benchmark ASD tug in head seas. *Ocean. Eng.* **2023**, *272*, 113800. [CrossRef]
18. Lv, Z.; Lv, H.; Fridenfalk, M. Digital Twins in the Marine Industry. *Electronics* **2023**, *12*, 2025. [CrossRef]
19. Mauro, F.; Kana, A. Digital twin for ship life-cycle: A critical systematic review. *Ocean Eng.* **2023**, *269*, 113479. [CrossRef]
20. Bertram, V. *Practical Ship Hydrodynamics*, 2nd ed.; Butterworth-Heinemann: Oxford, UK, 2012.
21. Carlton, J. *Marine Propellers and Propulsion*, 4th ed.; Butterworth-Heinemann: Oxford, UK, 2018.
22. Virtanen, P.; Gommers, R.; Oliphant, T.E.; Haberland, M.; Reddy, T.; Cournapeau, D.; Burovski, E.; Peterson, P.; Weckesser, W.; Bright, J.; et al. SciPy 1.0: Fundamental algorithms for scientific computing in Python. *Nat. Methods* **2020**, *17*, 261–272. [CrossRef] [PubMed]
23. Hunter, J. Matplotlib: A 2D Graphics Environment. *Comput. Sci. Eng.* **2007**, *9*, 90–95. [CrossRef]
24. Lindstad, H.; Eskeland, G.S.; Psaraftis, H.N. Maritime Shipping and Emissions: A Three-Layered, Damage-Based Approach. *Ocean Eng.* **2015**, *10*, 94–101. [CrossRef]
25. Chen, B.-Q.; Soares, C.G.; Videiro, P. Review of digital twin of ships and offshore structures. In *Developments in Maritime Technology and Engineering*; CRC Press: London, UK, 2021; p. 7. [CrossRef]

Journal of
Marine Science and Engineering

MDPI

Article

A Modified Body Force Model for a Submerged Waterjet

Dakui Feng, Yongyan Ma, Zichao Cai, Pengwei Yang and Yanlin Zou *

Key Laboratory of Ship and Ocean Hydrodynamics of Hubei Province, School of Naval Architecture & Ocean Engineering, Huazhong University of Science and Technology, Wuhan 430074, China; feng_dk@hust.edu.cn (D.F.); mayongyan@hust.edu.cn (Y.M.); m202372543@hust.edu.cn (Z.C.); m202472731@hust.edu.cn (P.Y.)
* Correspondence: zouyanlin@hust.edu.cn

Abstract: The submerged waterjet exhibits advantages such as uniform inflow, minimal flow distortion, and excellent acoustic performance, making it particularly suitable for high-speed vessels. This study investigates the open-water characteristics of the submerged waterjet and develops a body force model for the submerged waterjet propulsion system. First, under uniform inflow conditions, numerical simulations were performed using the body force method by replacing the rotor with a virtual blade and simultaneously replacing both the rotor and stator. The results of the body force model were then compared in detail with those obtained using the sliding mesh method. Second, the influence of the inflow velocity plane position on the results of the body force model was analyzed. The results indicate that the body force method, which replaces both the rotor and stator with a virtual blade, fails to accurately simulate the forces acting on various components of the propeller and the true distribution of the propeller's flow field. In contrast, the method that replaces only the rotor with a virtual blade produces results for component forces and flow fields that are largely consistent with the results of the sliding mesh method, demonstrating the stability and reliability of the body force model. Additionally, the position of the inflow velocity plane has no significant effect on the model's computational results.

Keywords: RANS; submerged waterjet; open-water performance; body force model

1. Introduction

Waterjet propulsion has become the preferred propulsion mode for high-speed and high-performance ships due to its advantages of high propulsion efficiency, good anti-cavitation performance, low underwater radiation noise, good maneuverability and no appendages outside the ship suitable for shallow water navigation [1]. Compared with the conventional tail plate waterjet propeller, the submerged waterjet propeller is installed on the bottom plate of the hull stern. The propeller is integrated in the groove at the bottom of the stern, and the nozzle is completely immersed in water. It has the advantages of uniform inflow, small flow deformation and good acoustic performance. In addition, submerged waterjet propulsion combines the characteristics and advantages of conventional propeller propulsion and stern-plate waterjet propulsion. It has important application value and good development potential, and is an important development direction of new propulsion technology for surface ships. Currently, research on waterjet propulsion primarily relies on two approaches: experimental model testing and numerical simulation.

In terms of experimental studies, Gong et al. [2] employed a vehicle-mounted three-dimensional underwater particle image velocimetry (PIV) system to measure the velocity distribution at the intake port during a self-propulsion test of a waterjet-driven ship model in a towing tank. Their findings indicated that the pumping action of the waterjet induced a transverse velocity vector on the cross-section, increasing inflow heterogeneity. Seo et al. [3] conducted a series of model tests to evaluate the performance of waterjet propulsion in a towing tank, including resistance tests, waterjet system tests, and self-propulsion tests, based on a high-speed ship dynamic prediction program.

Despite their reliability, ship model tests have several limitations, including high time and labor requirements, significant costs, constrained experimental conditions, and limited data acquisition capabilities. Computational fluid dynamics (CFD) offers an efficient and cost-effective alternative by enabling full-field flow data acquisition through numerical simulation. Moreover, CFD allows flexible simulation of various operating conditions, effectively overcoming the limitations of traditional experiments and providing more comprehensive technical support for ship design and optimization. With rapid advancements in CFD methodologies and computing power, numerical simulation has become the primary research tool, while model testing is increasingly used for validation purposes. Han et al. [4] numerically predicted the six-degree-of-freedom self-propelled maneuvers of the submarine model BB2 using the CFD method, including steady-state steering maneuvers and space spiral maneuvers. Based on the published test results of the tank model, they verified the effectiveness of the CFD steady-state steering maneuver prediction and conducted the space spiral maneuver prediction on this basis. Ma et al. [5] took the KRISO container ship as the research object, adopted the RANS equation, and respectively used the fluid volume method and the physical force method to study the hydrodynamic characteristics of the model-scale ship and the full-size ship passing through the limited water area at low speed. The results show that the variations of the model and the full-scale vessel in shock force, sway force and yaw moment are basically the same.

Several studies have leveraged CFD to analyze waterjet propulsion systems. Park et al. [6] conducted numerical simulations to investigate the complex three-dimensional viscous flow phenomena, including interactions among the intake duct, rotor, stator, and contraction nozzle. Donyavizadeh et al. [7] discussed the influence of stator blade curvature and thickness on the hydrodynamic performance of a waterjet propeller. Yi et al. [8] conducted resistance and self-propulsion tests on submerged waterjet propulsion ships and used numerical simulations to obtain force distribution data for each component and flow field information. Their results showed strong agreement between computed ship resistance and shaft thrust values and those obtained from experiments, validating the reliability of the numerical approach. Additionally, Cao et al. [9] compared the performance of a single jet pump under uniform inflow conditions with that of a complete jet propulsion system under non-uniform inflow conditions using steady-state simulations. Huang et al. [10] examined the effect of vortex generators on flow separation at the inlet of a waterjet propulsion system, demonstrating that the generators suppressed flow separation along the inlet ramp wall and improved flow uniformity before entering the pump.

Further developments in waterjet propulsion research include novel propulsion configurations and performance enhancements. Yang et al. [11] introduced a new type of waterjet propulsion system centered on a positive displacement pump, comparing its advantages and disadvantages against traditional negative displacement pump systems. Zhao et al. [12] used three-dimensional numerical simulations to study the hydrodynamic characteristics and flow fields of waterjet propulsion under mooring conditions. Cai et al. [13] proposed an integrated propulsion system combining the propeller, power source, and shafting, which was attached to the hull via thrust transmission, and developed a method for measuring thrust in waterjet propulsion. Carlo et al. [14] used computational fluid dynamics (CFD) to study the scale effect of open-water propeller performance. The results showed that the open-water characteristics had a significant scale effect. When the transition model was applied to the numerical simulation of the model scale, the scale effect was significantly reduced. Mikulec et al. [15] presented the verification and validation of full-scale ship trials and CFD simulations. In accordance with industry standards, speed/power trials were conducted under three different power settings. Full-scale unsteady RANS CFD simulations were performed. The results demonstrated good agreement between the numerical predictions and experimental data, confirming the applicability of the numerical method for evaluating full-scale ship performance.

The propeller body force model is an effective approach for evaluating propeller performance. Unlike conventional methods that require detailed geometric modeling and complex rotational motion simulations, the body force method introduces propeller forces as source terms in the momentum equation, simplifying modeling while maintaining high computational efficiency. This method enables accurate prediction of flow field characteristics and is widely applied in propeller propulsion studies. Wu et al. [16] utilized the body force method to predict hull self-propulsion performance, while Fu et al. [17] implemented it for ship self-propulsion calculations at model scale. Hao et al. [18] applied the body force method to simulate ship turning maneuvers. Liu et al. [19] proposed an inside-outflow coupling simulation strategy by integrating lift line theory with the body force method to analyze lift and drag characteristics in an aerodynamic propulsion coupling configuration. Marko et al. [20] introduced a modified body force method that extended its application to azimuth thrusters, addressing the traditional model's inability to account for lateral forces. Their results demonstrated that the modified approach effectively reproduced force and flow field characteristics while significantly reducing computational workload.

In recent years, with the increasing adoption of waterjet propulsion, research on body force methods for waterjet propulsion systems has expanded. Hino and Ohashi [21] compared and analyzed free-surface variations near the hull of a waterjet-propelled ship under light-body and propeller conditions. They used the body force method to examine the effects of waterjet propulsion and duct structures on the surrounding flow field. Takai et al. [22] applied the body force method to numerically simulate the self-propelled Joint High-Speed Sealift (JHSS) ship. Liu et al. [23] employed the body force method to investigate the influence of waterjets on a waterjet-propelled amphibious ship, comparing the method's predictions against experimental results and those obtained via the discrete propeller method. Their findings indicated that the body force method achieved higher prediction accuracy. Eslamdoost et al. [24] proposed three different body force models for waterjet propulsors. The first model geometrically removes the impeller and stator blades, applying only the axial forces generated by both the rotor and stator within the body force model. The second model adds tangential forces to the axial forces applied in the first model. The third model replaces only the impeller blades while retaining the stator blades, duct, and other components, applying both the axial and tangential forces generated by the rotor within the body force model. Furthermore, simulations were conducted using methods such as the sliding mesh method and the moving reference frame method. Comparisons of the results from these methods with those from the body force models demonstrated that the third model (retaining the stator and replacing the impeller blades with body forces) provided superior simulation performance. Additionally, compared to the sliding mesh method, the body force method significantly reduced computational time.

The development of a reliable body force model is crucial for the design and performance evaluation of submerged propulsion systems. While previous studies have primarily focused on applying the body force method to simulate ship self-propulsion, research on the construction and application of body force models specifically for waterjet propellers remains limited. In this study, a series of numerical investigations on a submerged waterjet propeller are conducted using CFD. An uncertainty analysis is performed to validate the numerical methodology. By analyzing the open-water characteristics of submerged jet

propulsion, a body force model tailored to submerged waterjet propulsion systems is developed. Additionally, the influence of inflow velocity plane position on the model's results is examined. The reliability of the proposed body force model is validated by comparing its results with those obtained from the sliding mesh method under uniform flow conditions.

2. Geometries Studied

To validate the research methodology, a type of propeller suitable for a submerged waterjet propulsion ship is employed, which is composed of duct, rotor and rear stator. The number of rotor blades is 5, and the number of rear stator blades is 7. The geometry model of the propeller is shown in Figure 1, the main parameters are shown in Table 1.

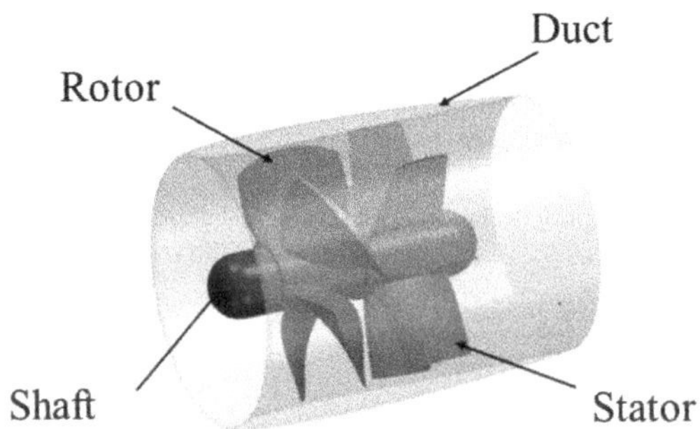

Figure 1. Geometry model of the waterjet propeller.

Table 1. Main parameters of the waterjet propeller.

Main Parameter	Symbol	Units	Value
Diameter of the rotor	D_R	m	0.166
Diameter of the Hub	D_{Shaft}	m	0.021
Hub diameter ratio	D_R/D_{Shaft}	-	0.25
Length of the duct	L_D	m	0.251
Inlet diameter	D_1	m	0.174
Outlet diameter	D_2	m	0.164
Blades number of rotor	N_D	-	5
Blades number of stator	N_R	-	7
Rotation direction	-	-	Right-handed

3. Computational Method

3.1. Governing Equation and Turbulence Model

In this study, numerical simulations of incompressible viscous flow are performed using star ccm+ 202302.0001. The CFD method is used to solve the unsteady RANS equations to simulate the flow characteristics of a submerged waterjet propulsion system. The governing equations of the URANS method are expressed as follows:

$$\frac{\partial(U_i)}{\partial x_i} = 0 \tag{1}$$

$$\frac{\partial U_i}{\partial t} + U_j\frac{\partial U_i}{\partial x_j} = -\frac{1}{\rho}\frac{\partial p}{\partial x_i} + \frac{1}{\rho}\frac{\partial}{\partial x_j}\left(\mu\frac{\partial U_i}{\partial x_j} - \rho\overline{u_i u_j}\right) \tag{2}$$

where U_i (i = 1, 2, 3) is the time-averaged velocity component and u_i (i = 1, 2, 3) refers to the fluctuation velocity component; ρ is the density; P is the time-averaged pressure; μ is the dynamic viscosity; and $\rho\overline{u_i u_j}$ is the Reynolds stress term.

To achieve turbulence closure, the shear-stress transport (SST) k-ω turbulence model, originally proposed by Menter [25], is employed.

The SST k-ω model enhances the accuracy of flow separation prediction and is well-suited for resolving complex flow fields. In the numerical simulations, a pressure-based coupling approach is employed for solving the governing equations. The convective terms are discretized using a second-order upwind scheme to enhance numerical stability and accuracy, while the diffusive terms are discretized using a second-order central difference scheme.

3.2. Body Force Model

The body force method is a special model that simulates the motion of a thruster, without real geometry, by adding the model at grid points within a given range to output force and torque. The application of the body force method can change the rotational motion of propeller from unsteady state to steady state. In addition, there is no need to divide the grid according to the real geometry, which greatly reduces the requirements on the mesh quantity and time step in the numerical simulation, greatly improves the computing speed and saves computing resources.

The body force method is designed to model the thrust, torque and other effects of the rotating part of the propeller, and it is not necessary to directly solve the motion of the actual rotating part. The application range of the body force region is a cylinder. This method distributes the body force f_b axially in the cylinder region, and follows a certain distribution in the radial direction. The distribution of force in this paper adopts Goldstein optimal distribution, and the relevant formulas are as follows:

$$f_{bx} = A_x r^* \sqrt{1 - r^*} \tag{3}$$

$$f_{b\theta} = A_\theta \cdot \frac{r^* \sqrt{1 - r^*}}{r^*(1 - r'_h) + r'_h} \tag{4}$$

$$r^* = \frac{r' - r'_h}{1 - r'_h},\ r'_h = \frac{R_H}{R_P}, r' = \frac{r}{R_P} \tag{5}$$

where f_{bx} and $f_{b\theta}$ represent the axial and tangential components of the body force, respectively. r is the radial coordinate; R_H is the radius of the hub; R_P indicates the radius of the tip circle. A_x and A_θ are constants, which can be defined as follows:

$$A_x = \frac{105}{8} \cdot \frac{T}{\pi\Delta(3R_H + 4R_P)(R_P - R_H)} \tag{6}$$

$$A_\theta = \frac{105}{8} \cdot \frac{Q}{\pi\Delta R_P(3R_H + 4R_P)(R_P - R_H)} \tag{7}$$

where T is propeller thrust; Q represents propeller torque; Δ indicates the thickness of the body force virtual disk.

3.3. Computational Domain and Boundary Condition

An appropriate computational domain measuring $15D_R$ (length) × $6D_R$ (diameter) is created for the propeller open-water simulations, the boundary conditions are shown in Figure 2. The velocity inlet is $5D_R$ upstream from the center of the rotor, and the pressure outlet is $10D_R$ downstream from the center of the rotor. The side of the computational domain is set as a symmetrical plane; the propeller surface is defined as a non-slip wall.

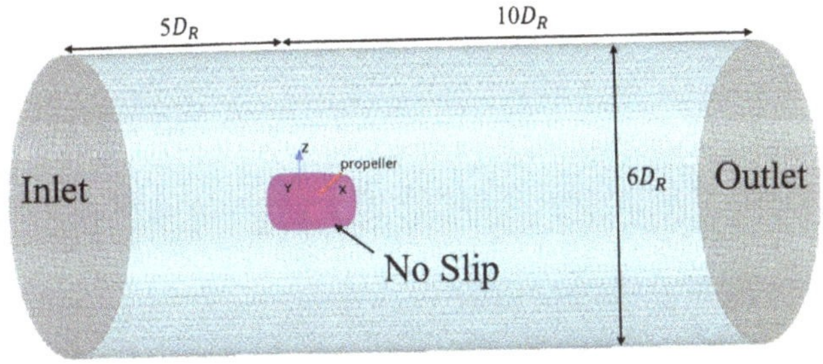

Figure 2. Computational domain and boundary condition.

The grid of the propeller open-water calculation is shown in Figure 3. To accurately simulate the flow field information during the rotation of the propeller rotor and meet the requirements of the selected turbulence model, the thickness of the near-wall mesh is set to 1.45×10^{-5} m, the number of boundary layers is not less than 16, as shown in Figure 3. The dimensionless wall distance Y+ distribution on the propeller surface is generally less than 1. The grid distribution is shown in Figure 4.

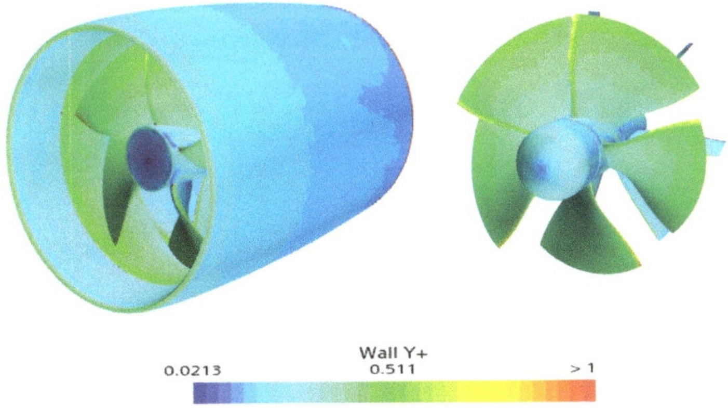

Figure 3. Y+ on the propeller surface.

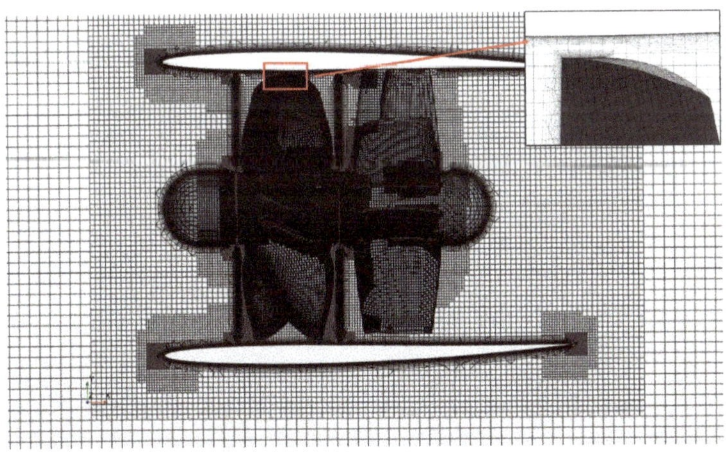

Figure 4. The grid of the propeller open-water calculation.

3.4. Numerical Validation

To verify the accuracy and reliability of the numerical method, based on the public test data, the uncertainty analysis is carried out by using the safety factor method proposed by Xing and Stern [26]. According to the regulations recommended by ITTC [27], the uncertainty of CFD numerical calculation can be expressed as

$$U_{SN} = \sqrt{{U_G}^2 + {U_T}^2 + {U_O}^2} \tag{8}$$

where U_G is the grid uncertainty; U_T is the time step uncertainty; and U_O represents the uncertainty caused by other factors, which can be ignored in subsequent analysis.

The refinement ratio r_k between solutions is assumed as

$$r_k = \frac{\Delta x_2}{\Delta x_1} = \frac{\Delta x_3}{\Delta x_2} \tag{9}$$

where Δx_i ($i = 1, 2$, and 3) are the coarse, medium, and fine grids or time steps, respectively.

According to the ITTC recommended procedures for uncertainty analysis in simulation calculations, the convergence coefficients R_k ($k = G$ denotes grid and $k = T$ denotes time step) are used to assess the convergence of grid and time step settings. The convergence factor R_k is defined as follows:

$$R_k = \frac{\varepsilon_{32}}{\varepsilon_{21}} \tag{10}$$

where $\varepsilon_{21} = \hat{S}_m - \hat{S}_c$, $\varepsilon_{32} = \hat{S}_f - \hat{S}_m$; $\hat{S}_f$, $\hat{S}_m$, $\hat{S}_c$ refer to the results of fine, medium and coarse, respectively.

Based on the convergence factor R_k for grid or time step, the convergence can be determined.

(1) $0 < R_k < 1$ Monotonic convergence (MC).

(2) $-1 < R_k < 0$ Oscillatory convergence (OC).

(3) $|R_k| > 1$ Divergence.

To conduct prediction analysis and validation of the hydrodynamic performance of a certain type of waterjet propulsion unit designed by the MARIC (Marine Design and Research Institute of China). The propulsion unit consists of a duct, a rotor, and a post-stator; the rotor has 6 blades, and the post-stator has 8 blades, with its geometric model shown in Figure 5.

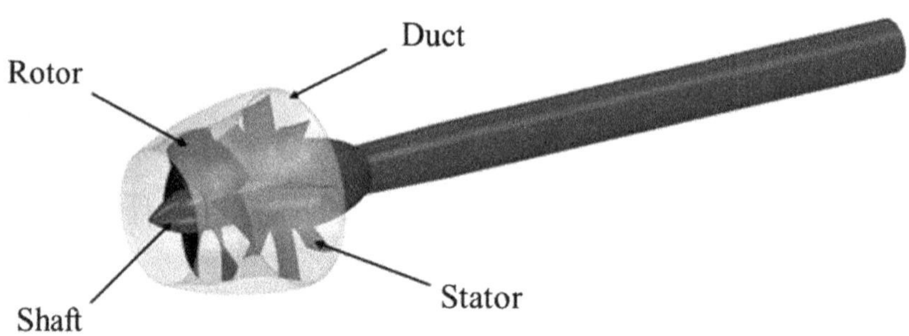

Figure 5. Geometry of a waterjet propeller.

Before conducting the waterjet simulations, the present numerical method has been verified and validated. In this study, the thrust coefficient for the propeller model is analyzed. The following coefficients are defined:

Advance coefficient:

$$J = \frac{V}{nD} \tag{11}$$

Thrust and torque coefficient:

$$K_T = \frac{T}{\rho n^2 D^4} \tag{12}$$

where V is the inflow velocity, n is the rotation speed, and T is the thrust.

The rotor thrust coefficient is K_T when the advance coefficient $J = 0.8$ is selected for numerical uncertainty analysis and calculation. The numerical results were compared with experimental fluid dynamics (EFD) data for the waterjet propeller, as presented in Table 2.

Table 2. Uncertainty analysis of a waterjet propeller.

	Grids Number (Million)	K_T	Error (%)	Time Step (s)	K_T	Error (%)
EFD	-	0.5608	-	-	0.5608	-
$\hat{S}_f$	11.891	0.5590	−0.32	1.26×10^{-4}	0.5601	−0.12
$\hat{S}_m$	8.696	0.5574	−0.61	2.53×10^{-4}	0.5574	−0.61
$\hat{S}_c$	6.537	0.5550	−1.03	5.05×10^{-4}	0.5512	−1.70
R_k	-	0.6974	-	-	0.4487	-
Convergence		MC			MC	

As shown in Table 2. The grid density and time step independence verification for K_T of the propulsor both showed monotonic convergence. The simulation results obtained using a medium grid size and a medium time step exhibited an error of −0.61% compared to the EFD data, which is within the numerical uncertainty range. This further validates the reliability of the numerical method employed in this study for predicting the hydrodynamic performance of the propeller.

For the submerged waterjet propeller studied in this paper, the advance coefficient J = 1.1 and rotational speed n = 16 rps are selected, with the thrust coefficient K_T as the target value for conducting sensitivity analyses of grid and time step size. A grid refinement ratio $r_k = \sqrt{2}$ and a time step refinement ratio $r_k = \sqrt{2}$ are chosen to estimate numerical uncertainty. The coarse, medium, and fine grids consist of 4.756 million, 6.432 million, and 8.706 million grid nodes, respectively. Time steps are set to 0.000694, 0.000347 and 0.000174 s, corresponding to 4, 2 and 1 degrees of propeller rotation per time step, respectively. The calculation results are shown in Table 3.

Table 3. Uncertainty analysis of waterjet propeller.

	Grids Number (Million)	K_T	Error (%)	Time Step (s)	K_T	Error (%)
Coarse	4.756	0.2968	1.55	0.000694	0.2958	1.23
Medium	6.435	0.2922	-	0.000347	0.2922	-
Fine	8.706	0.2897	−0.87	0.000174	0.2991	−0.11

The sensitivity analysis of grid and time step size is shown in Table 3. The computational results from three different grid density schemes and three different time step schemes demonstrate good convergence, further indicating the reliability of the numerical method used in this study for predicting the hydrodynamic performance of the propulsor.

Considering both computational accuracy and cost, the medium grid size and medium time step configuration are selected for subsequent calculations.

4. Construction of Body Force Model

4.1. Open-Water Characteristics of the Discretized Propeller

To analyze the hydrodynamic performance of the propeller, the sliding mesh method was employed to predict its hydrodynamic characteristics. This method requires direct grid discretization of the rotor geometry, allowing for an accurate representation of both the rotor's shape and its actual motion in the numerical simulation. Consequently, it effectively captures the real flow conditions within the flow channel.

The hydrodynamic simulation of the propeller was conducted using a fixed rotational speed approach while varying the incoming flow velocity. In this study, the propeller operated at a constant speed of n = 16 rps, with the incoming flow velocity V_A ranging from 0.2656 m/s to 3.4528 m/s. The time step for the rotor's rotational simulation was set to 0.000347 s, corresponding to a 2° rotation per time step at 16 rps.

The open-water performance results of the propeller are presented in Figure 6, where J represents the advance coefficient; K_{TR} is the rotor thrust coefficient; K_{TA} is the combined rotor and stator thrust coefficient; K_Q is the torque coefficient; and η denotes the propeller efficiency, defined as

$$J = \frac{V_A}{nD_R} \tag{13}$$

$$K_{TR} = \frac{T_R}{\rho n^2 D_R^4} \tag{14}$$

$$K_{TA} = \frac{T_R + T_S}{\rho n^2 D_R^4} \tag{15}$$

$$K_Q = \frac{Q}{\rho n^2 D_R^5} \tag{16}$$

$$\eta = \frac{JK_T}{2\pi K_Q} \tag{17}$$

where T_R is the thrust of propeller rotor, T_S is the stator thrust, ρ represents the fluid density, Q is torque, and rotor and stator thrust $T_A = T_R + T_S$.

4.2. Body Force Method

When constructing the body force model, the actual geometric structure of the propeller components does not need to be considered. Instead, the body force is assumed to be uniformly distributed along the axial direction on a virtual cylindrical disk. In this study, based on the open water results calculated by the sliding mesh method, a modified body force method is employed to analyze two scenarios: replacing only the rotor, and replacing both the rotor and stator.

The inlet section of the propeller duct is selected as the inflow velocity plane for the body force method. The incoming flow velocity V_A is corrected using the average flow velocity at the inlet of the propeller duct, V_{inlet}, to refine the open-water curve. The corrected advance coefficient, J_{inlet}, is defined as

$$J_{inlet} = \frac{V_{inlet}}{nD_R} \tag{18}$$

where n is the rotational speed and D_R is the rotor diameter.

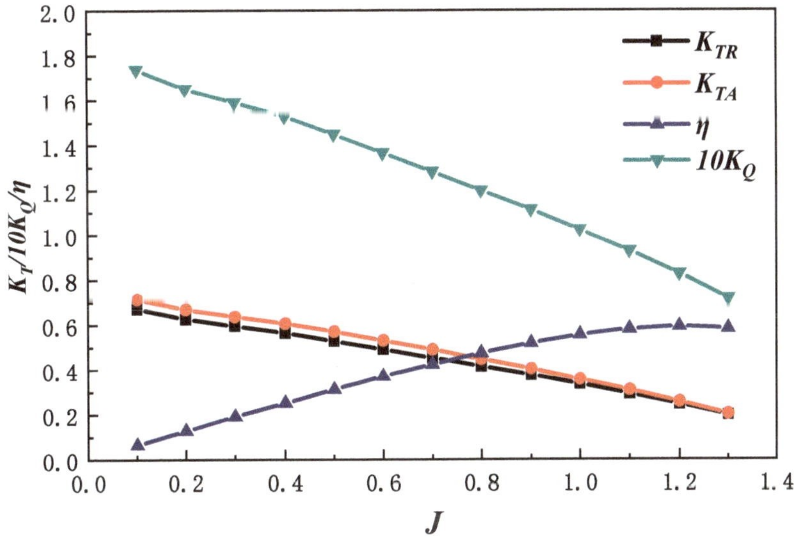

Figure 6. The open water characteristic curve of the propeller.

Figure 7 presents the average velocity distribution across the inlet section of the duct at different inflow velocities, while the modified open-water characteristic curve is shown in Figure 8.

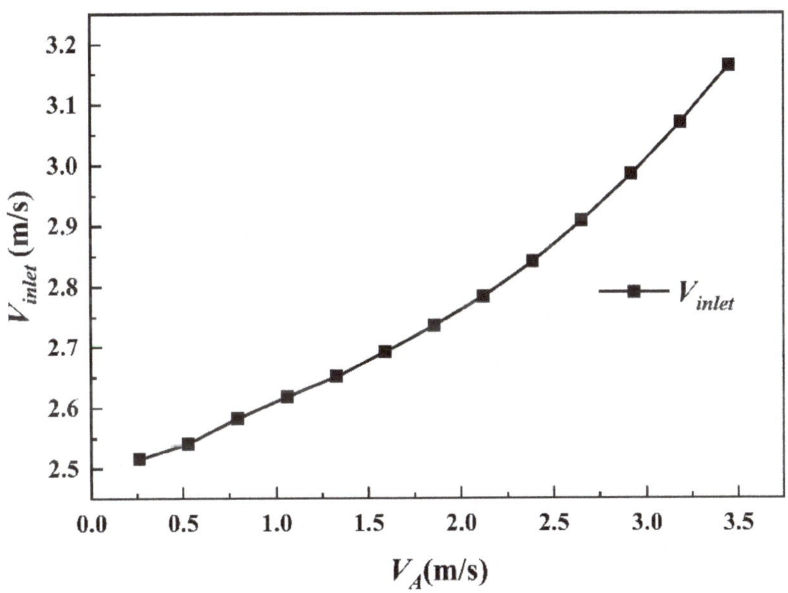

Figure 7. The mean cross-sectional velocity at inlet of the duct.

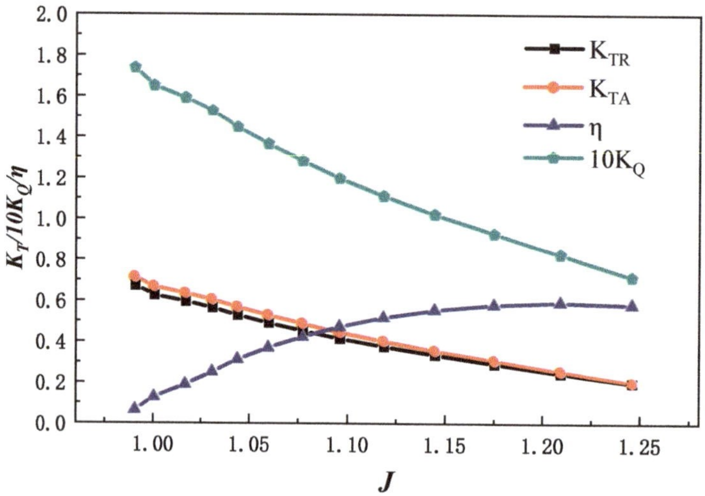

Figure 8. Modified open water curve.

5. Verification for the Results

5.1. Body Force Method Replaces Rotor Geometry

The rotor blade is replaced (referred to as AD-R) using the modified body force method described earlier, while keeping the shafting, stator blades, and duct intact. As shown in Figure 9, the applied body force region is modeled as a cylinder, where its diameter corresponds to the rotor blade diameter, and its thickness is the axial length of the rotor blade. The input for the body force method consists of J_{inlet}-K_{TR} and J_{inlet}-K_Q. The section at the inlet of the propeller duct is selected as the inflow velocity plane for AD-R.

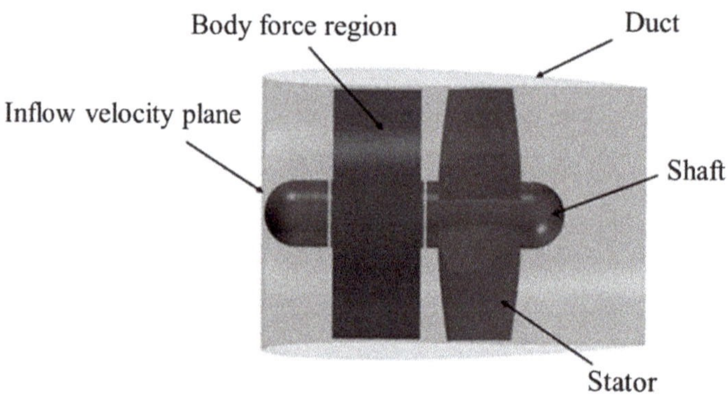

Figure 9. Illustration of AD-R.

For the simulation of the replacement rotor blade, the fixed body force speed n = 16 rps is consistent with the simulation of the sliding mesh method, with different flow speeds varied within the range of 1.062 m/s to 3.4528 m/s. The comparison of the thrust coefficient and torque coefficient of the propeller rotor calculated by AD-R and the sliding mesh method is presented in Table 4. Here, the numerical simulation results of the sliding mesh method are represented by subscript 0, the numerical simulation results of AD-R are represented by subscript 1, δ represents the parameter change rate of the former relative to the latter.

Table 4. Comparison of propeller performance between the sliding mesh method and AD-R.

V_A (m/s)	K_{TR1}	K_{TR0}	δ (%)	$10K_{Q1}$	$10K_{Q0}$	δ (%)
1.0624	0.5745	0.5646	1.754	1.5419	1.5283	0.889
1.3280	0.5360	0.5280	1.525	1.4604	1.4489	0.790
1.5936	0.5054	0.4903	3.094	1.3952	1.3661	2.129
1.8592	0.4632	0.4530	2.243	1.3048	1.2839	1.626
2.1248	0.4210	0.4146	1.548	1.2140	1.1983	1.311
2.3904	0.3763	0.3760	0.070	1.1169	1.1137	0.289
2.6560	0.3298	0.3352	−1.606	1.0149	1.0239	−0.878
2.9216	0.2823	0.2922	−3.410	0.9087	0.9296	−2.249
3.1872	0.2353	0.2465	−4.523	0.8011	0.8276	−3.194
3.4528	0.1921	0.1981	−3.043	0.6977	0.7178	−2.804

From Table 4, it can be observed that AD-R can accurately simulate the rotation of the rotor. The thrust coefficients calculated by both methods are within a 5% difference, and the error in the torque coefficient is less than 5%.

The forces of each component of propeller under the simulation of two methods are compared and analyzed. The comparison between the total thruster thrust calculated by AD-R and the sliding mesh method and the force on the conduit and other components is shown in Figure 10. Where T_{duct} represents the force of the propeller duct, and T_{total} represents the total thrust of the propeller.

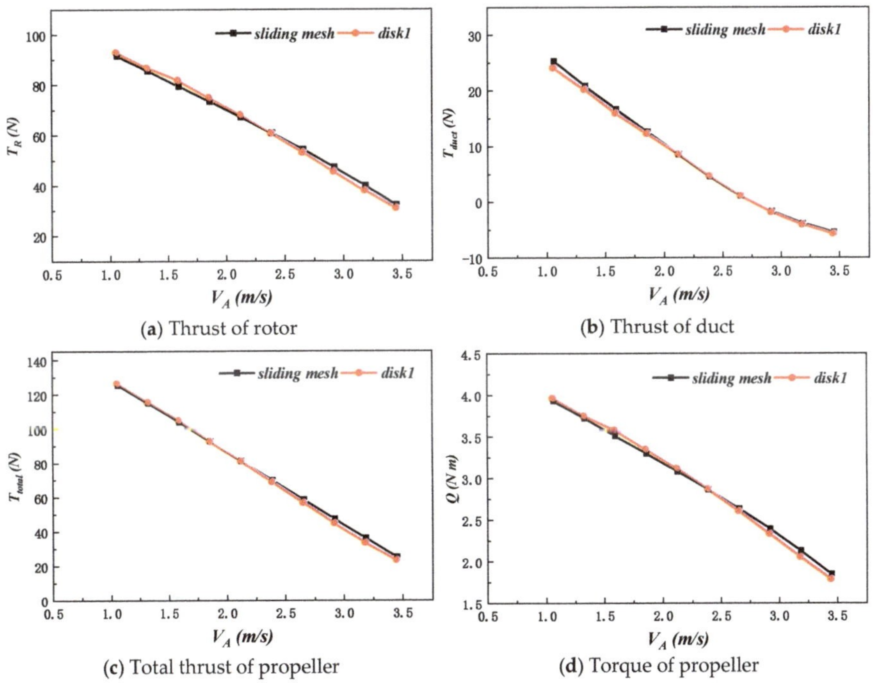

(**a**) Thrust of rotor (**b**) Thrust of duct (**c**) Total thrust of propeller (**d**) Torque of propeller

Figure 10. Comparison of propeller forces between the two methods.

As shown in Figure 10, AD-R can effectively simulate the forces on the entire propeller and its components. In most of the velocity range, the total thrust error between AD-R and the sliding mesh method is less than 5%. Additionally, the forces on the duct and rotor are in good agreement.

The flow field of AD-R during the simulation of propeller motion was also analyzed. A comparison of the flow field performance of the sliding mesh method and body force method was conducted at the same incoming flow velocity. When the flow velocity is 2.656 m/s, the comparison of the velocity distribution of the two methods at the same cross-section is shown in Figures 11 and 12. As seen in the figures, the longitudinal velocity distribution at the inlet and outlet sections of the duct is nearly identical for both methods.

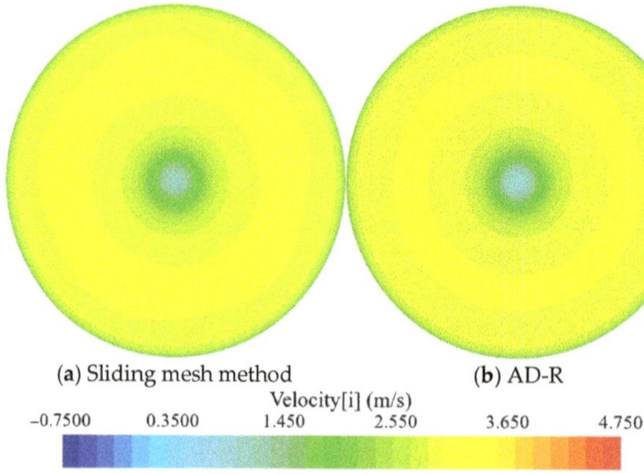

Figure 11. Comparison of velocity distribution of inlet section.

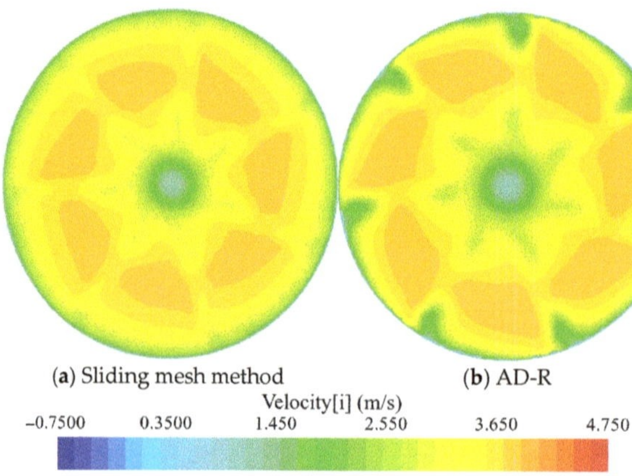

Figure 12. Comparison of velocity distribution of outlet section.

Additionally, the average axial velocity at both the inlet and outlet sections of the duct is calculated for different inflow velocities. The comparison between the two methods is shown in Table 5. Here, $V_{inlet\_}$ *AD-R* and $V_{outlet\_}$ *AD-R* represent the average axial velocities of the inlet and outlet sections of the duct, respectively, using the *AD-R* method. Meanwhile, V_{inlet} and V_{outlet} represent the average axial velocities at the inlet and outlet sections of the duct calculated using the sliding mesh method.

Table 5. Comparison of the average axial velocity of the inlet and outlet sections of the duct.

V_A (m/s)	V_{inlet1} (m/s)	V_{inlet0} (m/s)	δ (%)	$V_{outlet1}$ (m/s)	$V_{outlet0}$ (m/s)	δ (%)
1.062	2.553	2.619	−2.50	2.881	2.951	−2.39
1.328	2.603	2.651	−1.83	2.939	2.989	−1.70
1.594	2.646	2.692	−1.71	2.989	3.036	−1.57
1.859	2.705	2.736	−1.14	3.059	3.089	−0.98
2.125	2.764	2.784	−0.70	3.130	3.147	−0.56
2.390	2.825	2.841	−0.57	3.197	3.211	−0.44
2.656	2.892	2.908	−0.55	3.269	3.282	−0.42
2.922	2.970	2.984	−0.50	3.350	3.363	−0.38
3.187	3.061	3.071	−0.31	3.446	3.452	−0.18
3.453	3.170	3.165	0.17	3.561	3.550	0.33

Under open-water conditions, the inflow at the duct inlet of the propeller is significantly influenced by the duct. Since AD-R retains the geometric shape of the propeller duct, as seen in Figure 11. And Table 5, the velocity distributions at the duct inlet cross-section obtained by both methods remain highly consistent. Furthermore, across different inflow velocities, the maximum error in the average axial velocity at the duct inlet cross-section computed by AD-R, compared to that obtained by the sliding mesh method, is only −2.50%.

Different from the flow at the inlet of the duct, the outlet section of the duct is located at the end of the propeller, and the water flow at the outlet will be affected by the suction effect caused by the rotation of the rotor. At the same time, the existence of the rear stator will have a rectification effect on the fluid flowing through the rotating rotor, making the propeller outflow uniform. As shown in Figure 12 and Table 5, due to the rectification effect of the rear stator, the velocity distribution of the duct outlet section under the two methods is basically the same. Furthermore, under different inflow velocities, the maximum error between the average axial velocity of the duct outlet section calculated by AD-R and the sliding mesh method is only −2.39%, and the error gradually decreases with the increase of the inflow velocity.

Figure 13 shows the comparison of the longitudinal velocity distribution on the XY plane, as well as the pressure distribution on the inner tube and stator surface, calculated by both methods.

As shown in Figure 13, when the incoming flow velocity is 2.656 m/s, and the velocity fields calculated by both methods are nearly identical, indicating that AD-R can effectively simulate both the internal and external flow fields of the propeller. However, noticeable differences are present in the pressure distributions along the inner duct and stator surfaces between the two methods.

Firstly, the sliding mesh method accurately simulates the actual rotor rotation. The gap between the rotor and the duct inner wall, along with the flow field variations induced by the rotor's motion, creates distinct low-pressure regions near the rotor-adjacent duct walls. Additionally, the non-uniform flow field distribution downstream of the rotating rotor leads to localized high-pressure zones at the stator's leading edge.

In contrast, the AD-R method substitutes the rotor geometry with an equivalent disk of the same diameter. This approach fundamentally differs from the unsteady flow characteristics created by physical blade rotation. As a result, AD-R yields a more uniform pressure distribution along the virtual disk-adjacent duct walls and lower absolute pressures on the stator blades.

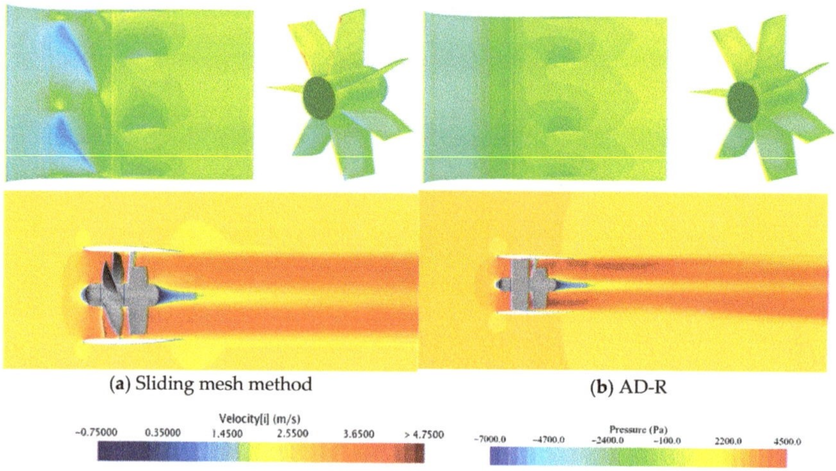

Figure 13. Comparison of propeller flow field calculated by two methods.

5.2. Body Force Method Replaces Rotor and Stator Geometry

To assess the applicability of the modified body force method for replacing both the rotor and stator blades, the body force model (designated as AD-RS) is used to replace the rotor and stator blades, while retaining the shafting and duct. The schematic configuration of AD-RS is shown in Figure 14.

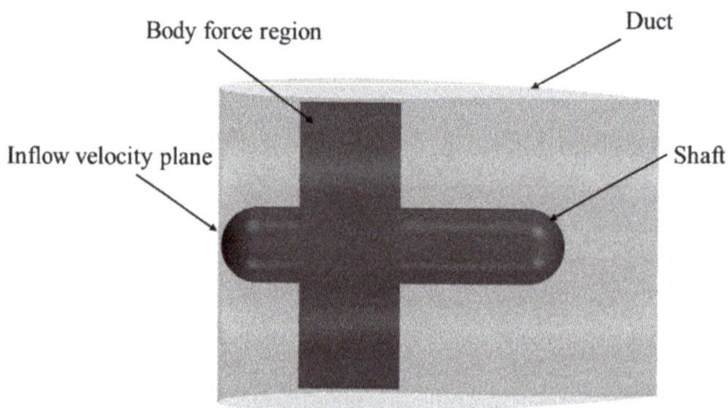

Figure 14. Illustration of AD-RS.

The rotational speed of the fixed body force is set to 16 rps, with a flow velocity range from 0.7968 to 3.4528 m/s. A comparison of the thrust coefficient and torque coefficient calculated by AD-RS and the sliding mesh method is provided in Table 6. The numerical simulation results of the sliding mesh method are represented by subscript0; the numerical simulation results of AD-RS are represented by subscript2.

As shown in Table 6, the thrust and torque coefficients of AD-RS are generally consistent with those of the sliding mesh method at lower flow velocities. However, as the flow velocity increases, the error between the two methods gradually increases.

To explore the reason of the large error in AD-RS at medium and high incoming flow velocities, the force and flow field characteristics of each propeller component simulated by both methods are compared. The comparison of the total thrust of the propeller and the forces on components such as the duct during the simulation calculations of AD-RS and the sliding mesh method is shown in Figure 15.

Table 6. Comparison of propeller performance between the sliding mesh method and AD-RS.

V_A (m/s)	K_{TA2}	K_{TA0}	δ (%)	$10K_{Q2}$	$10K_{Q0}$	δ (%)
1.062	0.5819	0.5646	−3.90	1.4734	1.5283	−3.59
1.328	0.5432	0.5280	−4.65	1.3950	1.4489	−3.72
1.594	0.5062	0.4903	−4.59	1.3201	1.3661	−3.37
1.859	0.4669	0.4530	−4.63	1.2410	1.2839	−3.34
2.125	0.4246	0.4146	−4.89	1.1560	1.1983	−3.53
2.390	0.3798	0.3760	−5.83	1.0664	1.1137	−4.25
2.656	0.3324	0.3352	−6.92	0.9725	1.0239	−5.02
2.922	0.2829	0.2922	−8.30	0.8750	0.9296	−5.87
3.187	0.2326	0.2465	−9.42	0.7774	0.8276	−6.06
3.453	0.1836	0.1981	−9.18	0.6844	0.7178	−4.65

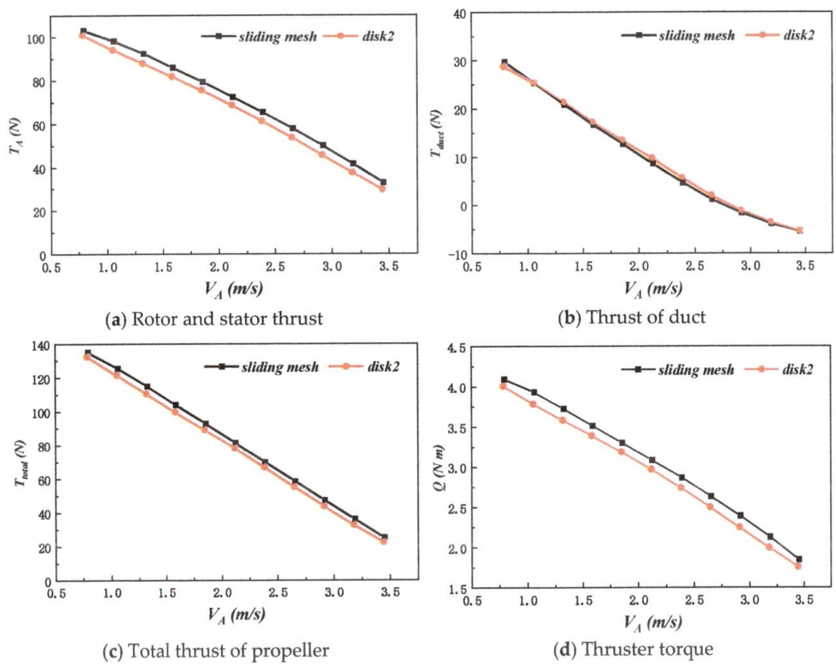

(**a**) Rotor and stator thrust (**b**) Thrust of duct (**c**) Total thrust of propeller (**d**) Thruster torque

Figure 15. Comparison of two methods of thruster force.

Based on the results in Table 6 and Figure 15, it is evident that the large error in the total thrust of the propeller is due to the significant discrepancies in the thrust and torque output of AD-RS at medium and high flow speeds. At an incoming flow velocity of 3.4528 m/s, the total thrust error of the propeller reaches −9.73%.

The comparison of the XY plane longitudinal velocity cloud diagram, pressure distribution on the inner tube surface, and velocity distribution at the outlet section of the duct calculated by AD-RS and the sliding mesh method is shown in Figure 16.

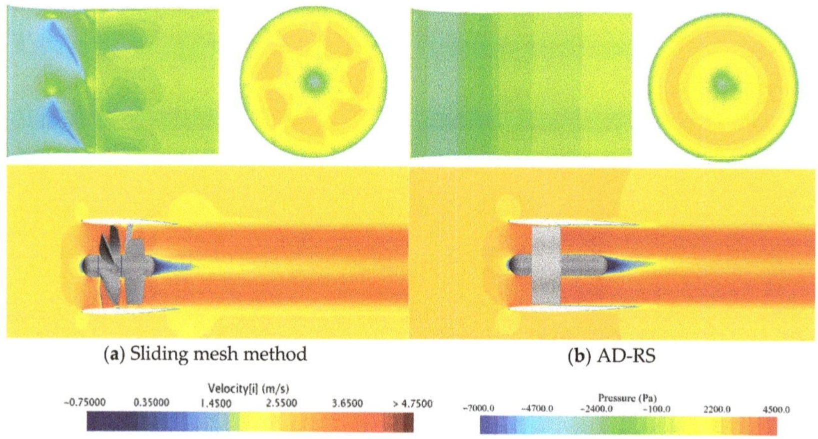

Figure 16. Comparison of the flow field of the propeller calculated by the two methods.

Both the rotor and stator blades are replaced geometrically in AD-RS, treating the rotor and stator as equivalent rotating components. Yet, this simulation cannot accurately replicate the effect of the rotor and stator on the flow field. As shown in Figure 16, since AD-RS replaces the stator's actual geometric blades in the simulation, the stator's rectifying action is entirely neglected. Consequently, the calculated flow field distribution at the conduit outlet does not match the actual flow field distribution at the thruster outlet, leading to an inaccurate simulation of the flow field inside the thruster. Furthermore, as the body force area in the AD-RS simulation is modeled as a cylinder and lacks the real rotor and stator blade geometries, the pressure distribution on the inner tube wall tends to be uniform across the entire section.

In conclusion, compared with AD-R, AD-RS treats the rotor and stator geometries as a unified rotating component, disregarding the stator's rectifying role in the propeller. This approach results in an inability to accurately simulate the forces on each component of the propeller, the true distribution of the flow field within the propeller, and the flow behavior at the propeller's outlet.

5.3. *Influence of Different Inflow Velocity Planes on Body Force Method*

Based on the analysis of the body force method replacing the rotor, the position of the inflow velocity plane is adjusted to study the influence of inflow velocity plane placement on the results of the body force method. As shown in Figure 17, the improved body force method is used to simulate inflow velocity planes at four different positions. Each disk surface is offset by $1\%D_R$, $5\%D_R$, $10\%D_R$ and $20\%D_R$ relative to the propeller disk surface. The radius of each inflow velocity plane is the radius of the inner duct at the longitudinal position of the plane.

The flow velocity V_A is modified using the average velocity V_{plane} at the inflow velocity plane, which further adjusts the open water curve. Figure 18 shows the section-averaged velocity V_{plane} distributions across the disk surfaces at different inflow velocities. As the offset increases from $1\%D_R$ to $20\%D_R$, the average cross-sectional velocity of the disk surface gradually decreases. This is due to the rotation and suction effect of the rotor, which accelerates the near flow field. Therefore, the closer the disk is to the propeller, the higher the average velocity at that disk.

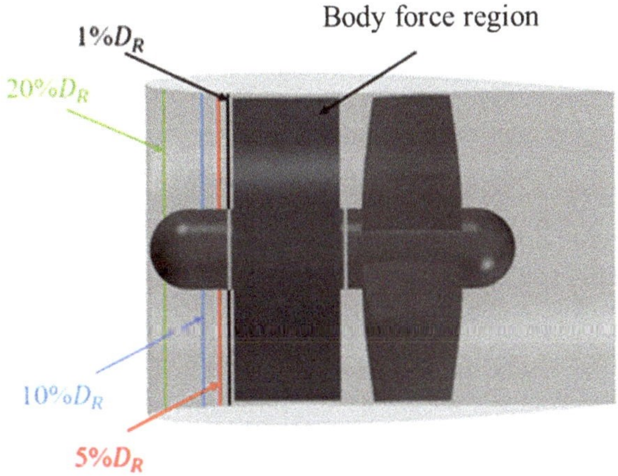

Figure 17. Schematic diagram of inflow velocity plane position.

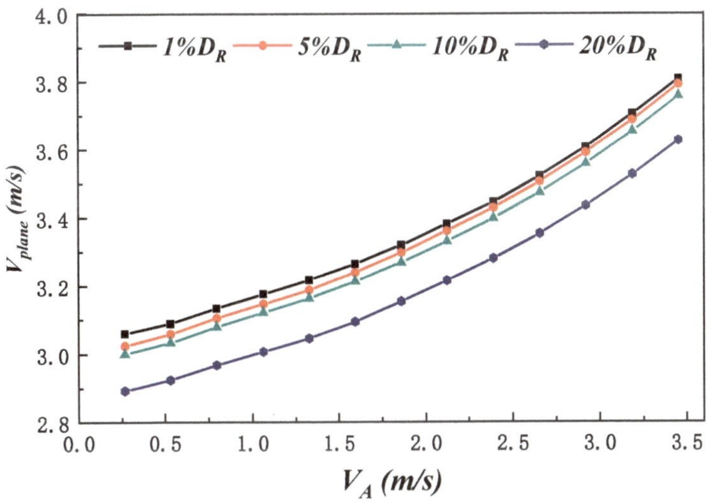

Figure 18. The average cross-sectional velocity of each disk surface.

Consistent with Section 5.1, the body force method is used to replace the rotor blades, and the shafting, stator blades and ducts are retained. The rotational speed of the fixed body force is 16 rps, and the velocity range is from 1.062 m/s to 3.453 m/s by changing the different incoming flow velocity V_A.

The open water characteristic curves corresponding to the four inflow velocity planes are input into the body force model to analyze the influence of the inflow velocity plane position on the results of the body force method. The hydrodynamic performance of the propeller is calculated for inflow velocity plane positions of 1%D_R, 5%D_R, 10%D_R and 20%D_R, respectively. The comparison of the thrust coefficient and torque coefficient of the propeller, as calculated by the body force method and the sliding mesh method, is shown in Table 7.

Table 7. Comparison of thrust coefficient and torque coefficient between two methods.

(a) Thrust Coefficient					
V_A (m/s)	K_{TR0}	K_T_1%D_R	K_T_5%D_R	K_T_10%D_R	K_T_20%D_R
1.062	0.5646	0.5889	0.5862	0.5877	0.5885
1.328	0.5280	0.5466	0.5458	0.5448	0.5476
1.594	0.4903	0.5066	0.5063	0.5048	0.5071
1.859	0.4530	0.4660	0.4668	0.4648	0.4683
2.125	0.4146	0.4259	0.4204	0.4254	0.4266
2.390	0.3760	0.3807	0.3800	0.3808	0.3822
2.656	0.3352	0.3351	0.3349	0.3357	0.3373
2.922	0.2922	0.2887	0.2889	0.2895	0.2913
3.187	0.2465	0.2420	0.2425	0.2429	0.2448
3.453	0.1981	0.1971	0.1971	0.1975	0.1995
(b) Torque Coefficient					
V_A (m/s)	K_{Q0}	K_Q_1%D_R	K_Q_5%D_R	K_Q_10%D_R	K_Q_20%D_R
1.0624	0.1528	0.1585	0.1582	0.1583	0.1586
1.3280	0.1449	0.1496	0.1492	0.1493	0.1496
1.5936	0.1366	0.1424	0.1420	0.1421	0.1424
1.8592	0.1284	0.1324	0.1330	0.1322	0.1326
2.1248	0.1198	0.1225	0.1225	0.1224	0.1228
2.3904	0.1114	0.1127	0.1126	0.1127	0.1131
2.6560	0.1024	0.1027	0.1027	0.1028	0.1033
2.9216	0.0930	0.0923	0.0925	0.0925	0.0930
3.1872	0.0828	0.0816	0.0819	0.0819	0.0825
3.4528	0.0718	0.0710	0.0712	0.0712	0.0718

It can be seen from Table 7 that in the numerical simulation, when the body force method employs different inflow velocity planes, the calculated thrust coefficient and torque coefficient are consistent with changes in the inflow velocity, Moreover, the error compared to the sliding mesh method remains consistently minor across all cases.

In order to analyze the accuracy of the body force method to simulate the open-water performance of the propeller when the inflow velocity plane at different positions is calculated, the force of each component of the propeller is compared and analyzed when the inflow velocity plane at different positions is calculated. The force on each component of the propeller is shown in Figure 19. It can be seen from the results in Figure 19 that when the inflow velocity planes at different positions are taken in the body force method, the force results of each component obtained by numerical simulation are not much different, and are in good agreement with the calculation results of the sliding mesh method.

Additionally, to analyze the flow field characteristics when the body force method is applied at different inflow velocity planes for propeller motion simulation, the flow field performance of the body force method after calculation stability is compared and analyzed at the same inflow velocity.

At different inflow velocities, the average axial velocity at each inflow velocity plane, calculated by the two methods, is shown in Figure 20. Here, $V_{plane-1\%D_R}$, $V_{plane-5\%D_R}$, $V_{plane-10\%D_R}$ and $V_{plane-20\%D_R}$ denote the average axial velocity at the inflow velocity plane with offsets of 1%, 5%, 10%, and 20%, respectively. The velocities calculated by body force method are in good agreement with those calculated by the sliding mesh method. The maximum errors are as follows: −1.77% for $V_{plane-1\%D_R}$, −2.2% for $V_{plane-5\%D_R}$, 2.16% for $V_{plane-10\%D_R}$, and 3.28% for $V_{plane-20\%D_R}$.

In addition, the average axial velocity of the duct outlet section of the propeller calculated by different inflow velocity planes is compared, as shown in Table 8. Here,$V_{ot-1\%D_R}$, $V_{ot-5\%D_R}$, $V_{ot-10\%D_R}$ and $V_{ot-20\%D_R}$ represent the average axial velocity of the duct outlet section with the inflow velocity plane offset of 1%D_R, 5%D_R, 10%D_R and 20%D_R, respectively.

It can be seen from Table 8 that the error between the average axial velocity of the duct outlet section of the propeller and the calculation result of the sliding mesh method is small when the inflow velocity plane is different.

Figure 21 shows the comparison of the longitudinal velocity distribution of the XY plane, the pressure distribution of the inner duct, and the velocity distribution of the outlet section of the duct calculated by body force method under different inflow velocity planes. The results indicate that the inflow velocity plane position has little effect on the flow field of the propeller.

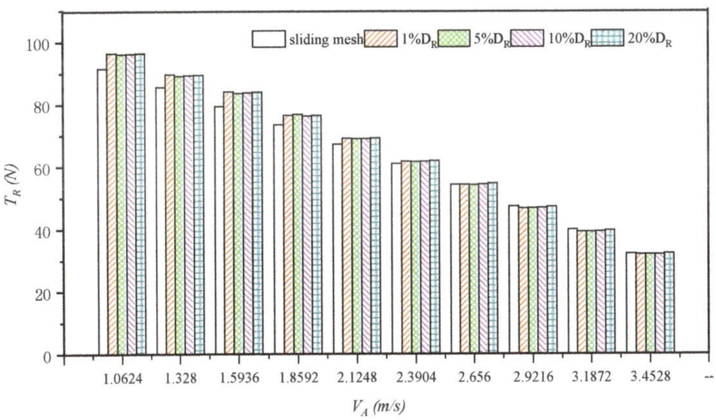

(**a**) Thrust of rotor

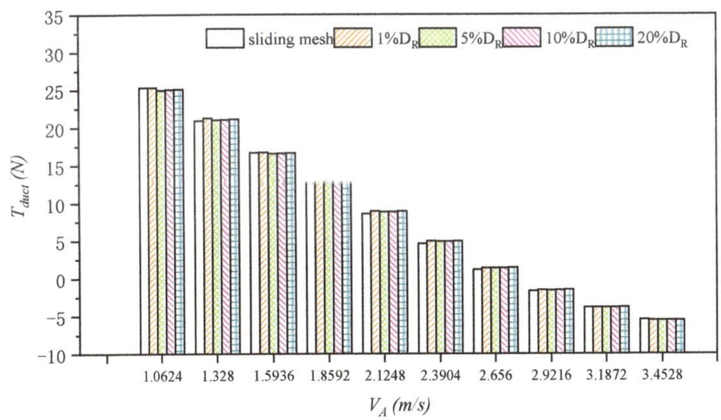

(**b**) Thrust of duct

Figure 19. *Cont.*

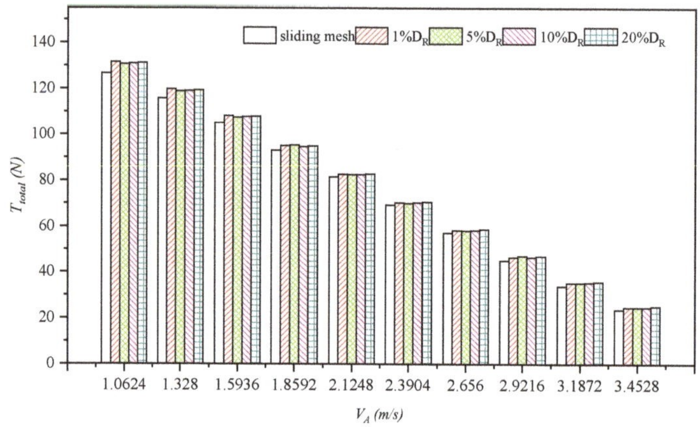

(**c**) Total thrust of propeller

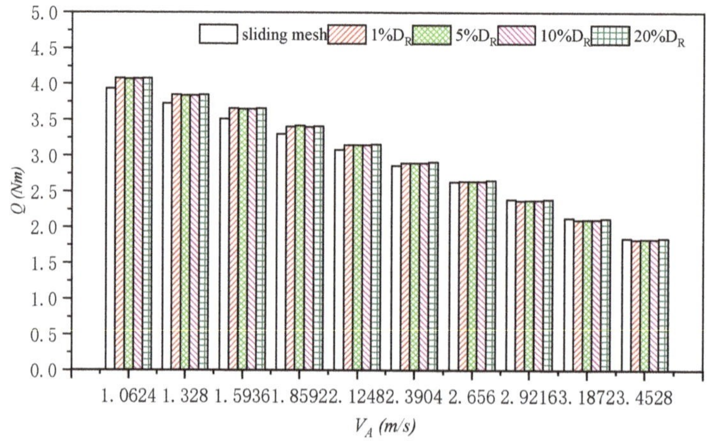

(**d**) Torque of propeller

Figure 19. The force of each component of the propeller in different disks.

Table 8. Average axial velocity of duct outlet section.

V_A (m/s)	$V_{outlet0}$ (m/s)	$V_{ot-1\%D_R}$ (m/s)	$V_{ot-5\%D_R}$ (m/s)	$V_{ot-10\%D_R}$ (m/s)	$V_{ot-20\%D_R}$ (m/s)
1.062	2.951	2.929	2.915	2.917	2.919
1.328	2.989	2.978	2.968	2.970	2.972
1.594	3.036	3.019	3.011	3.013	3.015
1.859	3.089	3.079	3.081	3.075	3.077
2.125	3.147	3.142	3.139	3.139	3.142
2.390	3.211	3.209	3.205	3.207	3.210
2.656	3.283	3.282	3.280	3.281	3.285
2.922	3.363	3.366	3.364	3.366	3.369
3.187	3.452	3.462	3.461	3.462	3.466
3.453	3.550	3.574	3.572	3.573	3.577

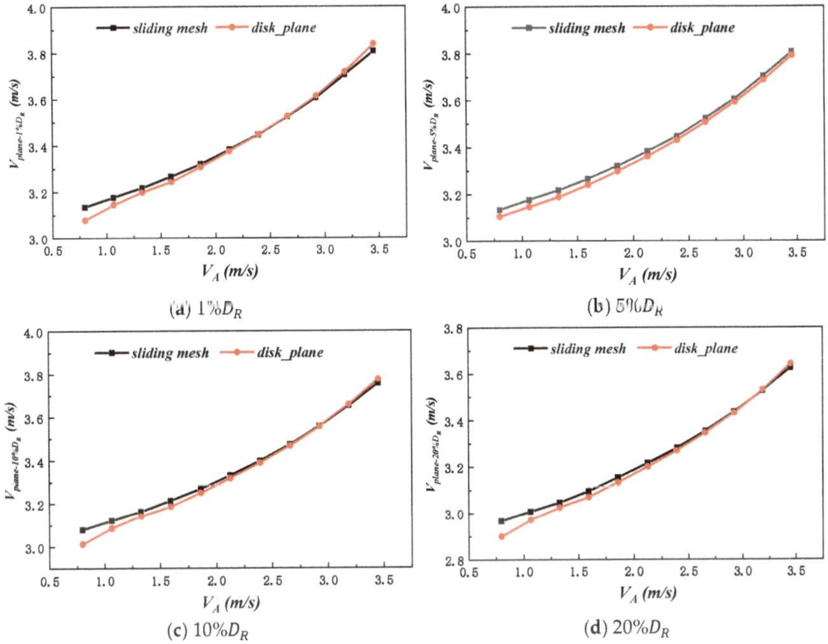

Figure 20. Comparison of average axial velocity distribution at different inflow velocity planes.

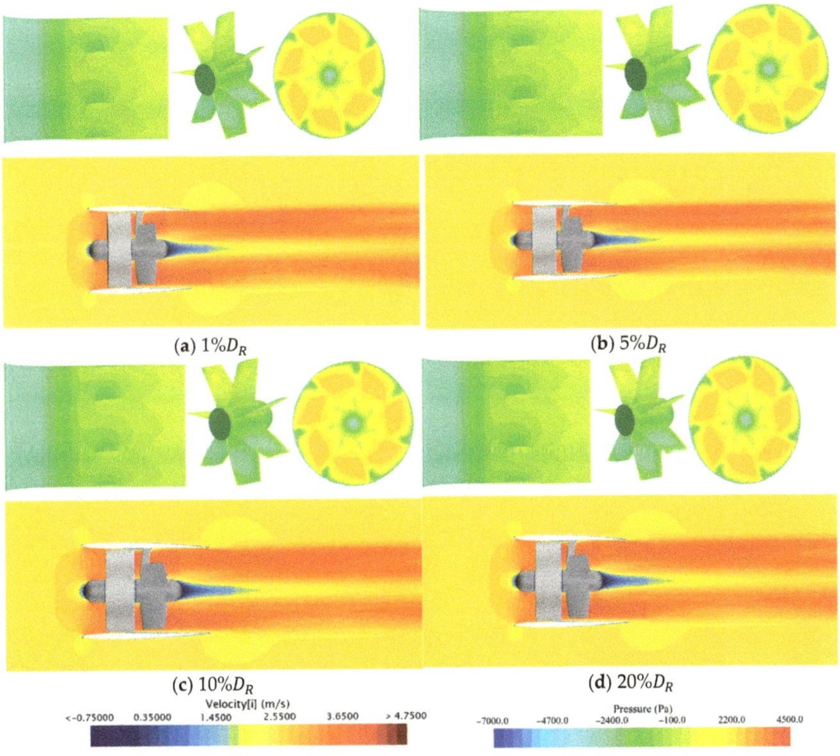

Figure 21. Comparison of propeller flow field under different inflow velocity planes.

In summary, when varying the positions of the inflow velocity plane, there is minimal impact on the calculation results of the improved body force method. Therefore, when simulating the propeller in open water, any plane between the duct inlet and the propeller disk surface can be chosen as the inflow velocity plane for the improved body force method. The average cross-sectional velocity at this plane can be used to correct the open water curve. This approach ensures the accuracy of the force on each component of the propeller when using the body force method for numerical simulations, and the flow field remains consistent with that calculated by the sliding mesh method.

6. Conclusions

This paper investigates the construction of the body force model for a submerged waterjet. A series of numerical simulations using the *RANS* method were conducted, and the numerical results showed good agreement with experimental data, verifying the reliability of the numerical method. A body force model suitable for submerged waterjet propulsion systems was developed, and the results were compared with those of the sliding mesh method under uniform inflow conditions. Additionally, the influence of the inflow velocity plane position on the results of the body force model was analyzed. The main conclusions are as follows:

(a) A body force model (AD-R) was developed to independently characterize rotor and stator forces. Under uniform inflow conditions, AD-R predicted rotor thrust and torque within 5% error compared to the sliding mesh method, with force distributions on propeller components and flow fields showing essential agreement.
(b) Compared with AD-R, AD-RS treats both rotor and stator geometries as a single rotating component, neglecting the stator's rectifying effect. Consequently, while yielding acceptable accuracy at low inflow speeds, this model exhibits significantly increasing errors with rising flow velocity. It is therefore unsuitable for simulating submerged waterjet propulsion systems.
(c) The influence of inflow velocity plane position on model accuracy was analyzed. Results demonstrate that the AD-R model maintains consistent performance across varying inflow plane locations, confirming its robustness for engineering simulations.

This study primarily focused on uniform inflow conditions, and future research can address the following areas: (1) optimizing the body force model to handle complex conditions such as non-uniform inflow and turbulence, and evaluating its applicability, and (2) studying the multi-field coupling effects between the propeller, hull, and other systems, and analyzing their influence on hydrodynamic performance. Such efforts would promote the broader application of the body force model in engineering practice and provide theoretical support for the design and optimization of submerged waterjet propellers.

Author Contributions: Conceptualization, D.F.; data curation, Y.M. and Y.Z.; formal analysis, Y.M. and Y.Z.; funding acquisition, D.F.; investigation, D.F.; methodology, Y.Z. and P.Y.; project administration, D.F.; resources, Y.Z. and Y.M.; software, Z.C. and P.Y.; supervision, D.F.; validation, Z.C.; writing—original draft, Y.M.; writing—review and editing, Y.M. and D.F. All authors have read and agreed to the published version of the manuscript.

Funding: This work was supported by the China National Science Foundation YEQISUN Joint Funds under grant U2141228.

Institutional Review Board Statement: Not applicable.

Informed Consent Statement: Not applicable.

Data Availability Statement: Data are contained within the article.

Conflicts of Interest: The research was conducted in the absence of any commercial or financial relationships that could be construed as potential conflicts of interest.

References

1. Allison, J. Marine Waterjet Propulsion. In Proceedings of the SNAME, Centennial Meeting, New York, NY, USA, 15–17 September 1993; Available online: https://trid.trb.org/view/444707 (accessed on 6 July 2025).
2. Gong, J.; Guo, C.; Wu, T.; Zhao, D. Particle image velocimetry measurement of velocity distribution at inlet duct of waterjet self-propelled ship model. *J. Hydrodyn. Ser. B* **2017**, *29*, 879–893. [CrossRef]
3. Seo, J.; Jeong, H.; Rhee, S.H.; Chang, K. Towing Tank Model Tests for Propulsive Performance Analysis of a Waterjet-Propelled Amphibious Vehicle. *J. Ship Res.* **2022**, *66*, 91–107. [CrossRef]
4. Han, K.; Cheng, X.; Liu, Z.; Huang, C.; Chang, H.; Yao, L.; Tan, K. Six-DOF CFD Simulations of Underwater Vehicle Operating Underwater Turning Maneuvers. *J. Mar. Sci. Eng.* **2021**, *9*, 1451. [CrossRef]
5. Ma, Z.; Ji, N.; Zeng, Q.; Deng, X.; Shi, C. Influence of scale effect on flow field offset for ships in confined waters. *Brodogradnja* **2024**, *75*, 75106. [CrossRef]
6. Park, W.; Jang, J.H.; Chun, H.H.; Kim, M.C. Numerical flow and performance analysis of waterjet propulsion system. *Ocean Eng.* **2005**, *32*, 1740–1761. [CrossRef]
7. Donyavizadeh, N.; Ghadimi, P. Efficacy Analysis of Thickness and Camber Size of Cross Section of the Stator on Hydrodynamic Parameters in Linear Jet Propulsion System. *Math. Probl. Eng.* **2020**, *2020*, 5861948. [CrossRef]
8. Yi, W.B.; Wang, Y.S.; Liu, C.J.; Peng, Y.L. Submerged waterjet self-propulsion test and numerical simulation. *Chuan Bo Li Xue/J. Ship Mech.* **2017**, *21*, 407–412.
9. Cao, P.; Wang, Y.; Kang, C.; Li, G.; Zhang, X. Investigation of the role of non-uniform suction flow in the performance of water-jet pump. *Ocean Eng.* **2017**, *140*, 258–269. [CrossRef]
10. Huang, C.; Dai, R.; Wang, Z. Effects of upstream vortex generators on the intake duct performance for a waterjet propulsion system. *Ocean Eng.* **2021**, *239*, 109838. [CrossRef]
11. Yang, Y.; Zhu, Y.; Luo, X. Water-jet Propulsion with Positive Displacement Pump. *Ship Eng.* **2011**, *33*, 31–34.
12. Zhao, D.G.; Zhang, Y.; He, Q.; Sun, C.; Bi, M.Q. Hydrodynamic and Flow Field Characteristics of Water Jet Propulsion under Mooring Conditions. *J. Mar. Sci. Eng.* **2022**, *10*, 953. [CrossRef]
13. Cai, Y.; Ji, G.; Feng, C.; Wang, Z.; Wang, L. Thrust Measurement Method of Water Jet Propulsion. *Shipbuild. China* **2022**, *63*, 107–114.
14. Grlj, C.G.; Degiuli, N.; Farkas, A.; Martić, I. Numerical Study of Scale Effects on Open Water Propeller Performance. *J. Mar. Sci. Eng.* **2022**, *10*, 1132. [CrossRef]
15. Mikulec, M.; Piehl, H. Verification and validation of CFD simulations with full-scale ship speed/power trial data. *Brodogradnja* **2023**, *74*, 41–62. [CrossRef]
16. Wu, Z.H.; Chen, Z.G.; Dai, Y. Numerical prediction of self-propulsion with a body-force propeller model. *J. Shanghai Jiaotong Univ.* **2013**, *47*, 943–949.
17. Fu, H.P.; Michael, T.J.; Carrica, P. Computation on self-propulsion at ship point based on a body-force propeller. *J. Ship Mech.* **2015**, *19*, 791–796.
18. Hao, H.; Chen, W.; Du, Y.; Zhang, Q. Numerical simulation of a ship turning motion based on body-force method. *Ship Sci. Technol.* **2025**, *47*, 43–48.
19. Liu, H.; Zhao, X.; Zhou, F.; Feng, Y.; Wang, Y. Aero-propulsion analysis of distributed ducted-fan propulsion based on lifting-line driven body-force model. *Chin. J. Aeronaut.* **2025**, *38*, 103126. [CrossRef]
20. Mikulec, M.; Piehl, H. Development of a modified body force method for modelling azimuth thrusters in CFD-based manoeuvring simulations. *Ocean Eng.* **2025**, *319*, 120171. [CrossRef]
21. Broglia, R.; Dubbioso, G.; Durante, D.; Mascio, A.D. Simulation of turning circle by CFD: Analysis of different propeller models and their effect on manoeuvring prediction. *Appl. Ocean Res.* **2013**, *39*, 1–10. [CrossRef]
22. Takai, T.; Kandasamy, M.; Stern, F. Verification and validation study of URANS simulations for an axial waterjet propelled large high-speed ship. *J. Mar. Sci. Technol.-Jpn.* **2011**, *16*, 434–447. [CrossRef]
23. Liu, B.; Xu, X.; Pan, D.; Wang, X. Research on shipping energy-saving technology: Hydrofoil amphibious vehicle driven by waterjet propulsion. *J. Clean. Prod.* **2023**, *382*, 135257. [CrossRef]
24. Eslamdoost, A.; Vikström, M. A body-force model for waterjet pump simulation. *Appl. Ocean Res.* **2019**, *90*, 101832. [CrossRef]
25. Menter, F.R. Two-equation eddy-viscosity turbulence models for engineering applications. *AIAA J.* **1994**, *32*, 1598–1605. [CrossRef]

26. Stern, F.; Wilson, R.V.; Coleman, H.W.; Paterson, E.G. Comprehensive Approach to Verification and Validation of CFD Simulations—Part 1: Methodology and Procedures. *J. Fluids Eng.* **2001**, *123*, 793–802. [CrossRef]
27. ITTC. Uncertainty Analysis in CFD Verification and Validation, Methodology and Procedures. In *ITTC Recommended Procedures and Guidelines*; No.7.5-03-01-01; International Towing Tank Conference (ITTC): Fukuoka, Japan, 2008.

Journal of
Marine Science and Engineering

MDPI

Article

Two-Phase Multi-Point Design Exploration of Submerged Nacelles for Marine Propulsive Pump Installation

Filippo Avanzi *, Andrea Magrini † and Francesco De Vanna †

Department of Industrial Engineering, Università degli Studi di Padova, 35131 Padova, Italy; andrea.magrini@unipd.it (A.M.); francesco.devanna@unipd.it (F.D.V.)

* Correspondence: filippo.avanzi@unipd.it

† These authors contributed equally to this work.

Abstract: Outboard Dynamic-inlet Waterjets (ODW) are axisymmetric units, powered by a self-contained pump, that, by processing a uniform undisturbed streamtube, can operate more efficiently than conventional marine propulsors. This feature also provides methodological convenience, enabling accurate numerical investigations of the system alone using 2D axisymmetric models. Leveraging this property, the present study bridges the gap on the design principles required to tailor ODW geometries across multiple operating conditions. Reynolds-Averaged Navier Stokes (RANS) equations are solved, including turbulence and cavitation models, to draw the propulsor's characteristic maps and identify two relevant operating points, set by the combination of a specified pump rotational regime with an advancing velocity. Simulations for these in- and off-design conditions are systematically performed over a database of 512 randomly sampled geometric variants. The corresponding results show that optimised shapes improving the inlet Pressure Recovery (PR) and nacelle drag at cruise conditions result in beneficial outcomes also at take-off operations, where lip cavitation may occur. Thus, analysing together the off-design PR and the cruise net force underscores their conflicting behaviour. In fact, while nacelles shortened by 12% can reduce overall drag and enhance nominal net thrust by 2%, designs featuring a 34% wider capture area improve off-design PR by over 1.5%, albeit at the cost of compromised propulsive efficiency under any operating range.

Keywords: multi-point nacelle optimisation; two-phase design exploration; outboard dynamic-inlet waterjet; optimal hydrodynamic geometries; inlet cavitation

1. Introduction

Although conventional propellers dominate the marine propulsion market, waterjet units represent a more convenient solution for specific applications [1]. In fact, thanks to improved efficiency above 30 knots, higher manoeuvrability, and noise mitigation capabilities, they result in a proper choice for military installations [2]. Additionally, despite sharing the presence of a bladed component, required to generate a momentum unbalance, waterjets leverage diffuser inlets that delay rotor tip cavitation, enabling safer operations at high advancing speeds [3,4]. Therefore, the intake is a distinctive component, providing direct control on the pump Net Positive Suction Head (NPSH), consequently affecting the efficiency of the entire propulsion system [5]. As a result, the inlet shape characterises the specific application and is commonly classified into flush and podded types, with the flush configuration representing the standard arrangement in current applications [2]. Extensive studies have been dedicated to this component to identify its typical head losses mechanisms. Among others, up to a 20% efficiency drop can be determined by: hull boundary layer ingestion and distortion, lip flow turning and evolution through bent ducts [6]. Thus, parameterising the inlet walls curves, Computational Fluid Dynamics (CFD) studies were adopted, combined with Design Of Experiments (DOE) and genetic algorithms, to mitigate detrimental effects and improve the system's efficiency [5,7]. Optimisation of waterjet pumps is another common research theme [8–10].

Sharing common features with both ducted propellers and waterjets, pumpjet units are an alternative solution, usually applied to underwater vehicles. They comprise both rotating and stationary blades, shrouded by axisymmetric nacelles, and are generally installed astern, enabling an efficient exploitation of the boundary layer generated along the tailcone [11,12]. Different configurations have been devised over the years leveraging slant shafts, including scoop inlets [13] and fully-outboard devices [14]. Although providing improvements, these solutions are not capable of overcoming inherent propulsive limits, making optimisation through shape parameterisation necessary [15–17].

In recent years, with the introduction of Outboard Dynamic-inlet Waterjets (ODW), industrial technology has paved the way for innovation in marine propulsion [18]. In fact, the possibility of operating isolated from the ship, thanks to electric integrated engines, allows this novel propulsor to process uniform mass flow rates, aligning thrust with the advancing direction, similarly to aeronautical propulsion principles. Current related research is limited to numerical investigations, where 2D axisymmetric models [19] and 3D installed configurations [20] were adopted to measure relevant geometric parameters and characterise the system's performance. Recent studies [21] performed detailed analyses on the intake shape to derive proper forebody design criteria for optimal pump integration. However, that investigation limited to cruise operations and no two-phase phenomenon was included.

At the present stage, substantial advancements are bounded by the lack of public experimental data to serve as a reference for validation. Therefore, the only available numerical model relies on equivalent configurations obtained from a combination of public geometries [22–24]. To this end, persistent efforts from the academic environment on the topic are expected to stimulate experimental research, potentially leading the way toward achievements comparable to those obtained for conventional marine propulsors. In fact, over the years, extensive testing has been conducted to accurately measure the performance of stationary [6,22,25] and rotating components [26–28] of waterjets and pumpjet units [29–32]. Building upon these reference measurements, reliable CFD methods could be developed, enhancing the predictive accuracy of numerical models. Particularly, this process has enabled the improvement of closure techniques applied to Reynolds-Averaged Navier Stokes (RANS) equations, making them the preferable approach for a wide range of applications. Offering robust and rapid predictions without excessive computational costs, the development of accurate turbulence and cavitation models [33–37] has established the role of RANS techniques as a standard within both the industrial and academic hydrodynamics communities.

In an effort to advance current knowledge on ODW, this study emphasises the need for a dedicated investigation into the design principles governing optimal nacelle shaping. The analysis aims to establish design criteria by identifying correlations between fundamental geometric features and the resulting propulsive performance. Particular focus is given to ensuring efficient operation across varying conditions while delivering a clean flow to the turbomachinery, with sufficient cavitation margin. To this end, critical points in the propulsor's characteristic map are first identified by selecting an appropriate pump rotational speed, enabling the assessment of multi-point performance across different geometries. A database of geometric variants is then generated and modelled using two-phase RANS simulations at both cruise and take-off operations, the latter accounting for the potential onset of internal lip cavitation. From the resulting CFD solutions, relevant propulsive metrics are extracted using Thrust-Drag Bookkeeping (TDB), a technique typically adopted in aero-engine research [38–40] and particularly efficient for ODW characterisation [21]. The collected statistics are subsequently analysed to disclose correlations with principal geometric parameters, highlighting how local wall profiles should be shaped to enhance specific propulsive performance. In summary, the study highlights the relevant design variables for the geometric specification of ODW nacelles, setting the basis to draw design groundrules for efficient propulsive pump installed operations across mutliple operating conditions.

The paper is organised as follows: Section 2 introduces the reference model and related computational methods. In Section 3, optimisation results are presented starting from the propulsor's characteristic maps, while in Section 4 conclusions are drawn.

2. Computational Methods

2.1. Reference Geometry

The reference ODW geometry for the present study is reported in Figure 1. Due to the limited availability of data from existing units, the outline derives from a manipulation of the waterjet ram inlet presented by Sobolewski [22], adapted to house the AxWJ-2 pump designed by Michael et al. [24]. The details regarding the integration strategy of the two components are discussed in a previous work by the authors [20], while Avanzi et al. [21] presented a systematic investigation, characterising the geometrical parameters relevant to the hydrodynamic design of the propulsor's intake.

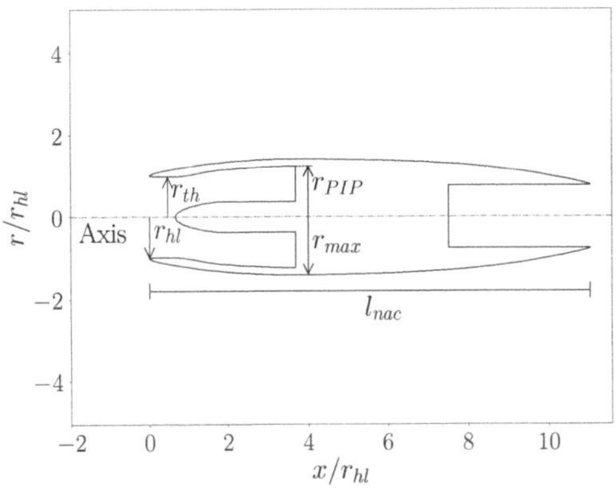

Figure 1. Meridional view of the baseline geometry of the Outboard Dynamic-inlet Waterjet (ODW).

This study extends the focus to the full system dynamics. Thus, a generic B-spline is adopted to complete the external nacelle geometry, connecting the outer inlet skin with the nozzle inner walls while preserving tangency continuity at the two sides. This curve parameterisation is convenient for local manipulation, providing a piecewise polynomial definition that preserves C^2 support on the whole shape [41]. For simplicity, the exhaust section comprises only the straight duct downstream of the pump stator. The omitted section corresponds to the casing of the machine stationary blades row, whose cross-sectional area distribution is determined by the profiles of the shroud and the spike.

The machine block is considered fixed, providing operations with a known characteristic map. Consequently, the sections across the pump model inflow and outflow conditions, keeping the Pump Interface Plane (PIP) and the exhaust planes unchanged. According to the chosen model, the intake modification still plays a pivotal role in the determination of the capture streamtube, while nozzle geometry manipulation is mainly considered effective for drag behavior. Specifically, the highlight (*hl*) and throat (*th*) sizes affect the quality of the entering flow, with the former being directly connected with the concept of Inlet Velocity Ratio (IVR). This parameter provides a quantitative description of the capture streamtube shape, following its definition:

$$IVR = \frac{Q}{A_{hl} v_\infty} \tag{1}$$

where Q is the volumetric flow rate through the ODW, A_{hl} is the highlight area, which is inherently related to the highlight radius, r_{hl}, and v_∞ is the free-stream velocity.

Quantifying the flow capacity of the system relative to undisturbed conditions, the IVR establishes a deterministic criterion for identifying the operating regime. Specifically, when $IVR > 1$, the parameter indicates a convergent pre-entry streamtube. In this case, the flow region is bounded by a stagnation line that reaches the intake wall at a point on the outer surface, shifted downstream from the optimal position at the highlight lip. As a result, the entering flow experiences higher accelerations compared to nominal operations. Differently, when $IVR < 1$, the opposite configuration arises-known as a pre-diffusion streamtube-which induces positive velocity gradients along the internal intake surface.

2.2. Numerical Setup and Boundary Conditions

A two-dimensional grid is adopted for the computational domain (Figure 2), assembling multiple structured blocks with a tessellation of quadrilateral elements. The far field boundary is separated at 100 r_{hl} from the near-wall region (Figure 2a) and is shaped as a half circle. The cells dimension is regulated with a geometric progression, starting from a first off-the-wall normal spacing lower than 10^{-5} r_{hl} to ensure an estimated $y^+ < 1$. The element size increases toward the free-stream boundaries, thereby limiting the overall mesh size while maintaining adequate average cell quality. Along wall boundaries, a denser nodes distribution is chosen near rounded edges (Figure 2b), gradually enlarging over straight outlines. Across the different refinements, the maximum included angle is monitored to never exceed 155°.

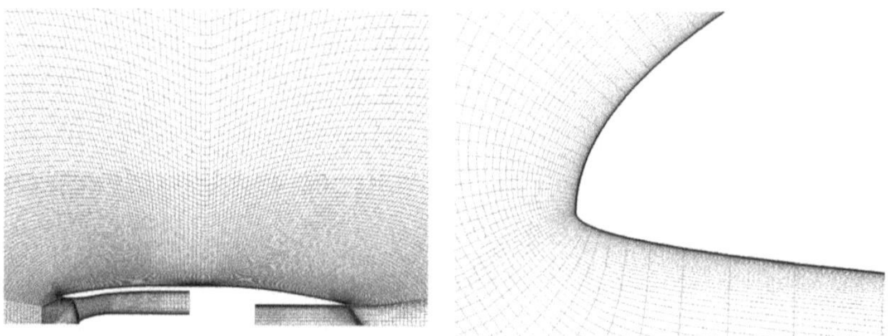

(a) (b)

Figure 2. Multi-block structured domain discretisation from near-wall region (**a**) and highlight leading-edge closeup view (**b**).

Steady-state RANS equations are formulated in 2D axisymmetric coordinates and solved using the pressure/velocity coupled solver Ansys Fluent® [42]. Continuity and momentum conservation laws are closed with the $\kappa - \omega$ Shear Stress Transport (SST) [43] and Zwart [44] models, to account, respectively, for turbulence and cavitation phenomena. A single volume fraction transport formulation is solved, coupled with the main system of governing equations, adopting a homogeneous mixture model. Second-order discretisation is adopted for momentum and turbulence terms, while volume fraction is discretised with a Quadratic Upstream Interpolation for Convective Kinematics (QUICK) approach. For pressure terms treatment, a PREssure STaggering Option (PRESTO!) scheme is chosen based on its suitability for cavitation modelling [45].

During initial iterations, mass flow rate is imposed as both inlet and outlet conditions, respectively at the PIP and the stator exit plane. Then, for two-phase simulations, the PIP boundary condition is switched to a pressure outlet to prevent the selection of the nature of the flow exiting the domain, enabling mixed-phase evolution along the connector. To mimic the "Target Mass Flow Rate" option available inside the solver only for single-phase flows, a custom approach is adopted to provide the pressure value at the boundary according to the unbalance between the current and expected mass flow rate. The implemented expression follows the definition reported in the solver's User Guide [42] for the built-in option. Concerning the farfield boundary, the connector is split into two equal arcs: the forward edge is set as velocity inlet, enforcing free-stream conditions, while the downstream half imposes ambient pressure through pressure outlet formulation. The propulsor's walls boundaries are treated as no-slip walls.

The solution initialisation is performed using a hybrid technique. Convergence strategies differ between single-phase and two-phase simulations. In the first case, the computations are generally smoother; therefore, a preliminary stage with first-order discretisation is sufficient to stabilise the solution. This is then followed by second-order discretisation until the convergence criteria are satisfied. These include monitoring the residuals of the governing equations, which must drop below $1 \cdot 10^{-5}$. Additionally, the x-component of the wall forces, summed over all body boundaries, is tracked, and its variation between two successive iterations must fall below $1 \cdot 10^{-6}$ to be considered converged.

For two-phase simulations, a dedicated strategy is employed to ensure convergence. The initial iterations are run using a first-order mixture-fluid arrangement, with phase-change phenomena temporarily deactivated. Once these computations stabilise the solution, the cavitation model is enabled, and a first-order discretisation is applied for a first set of iterations before transitioning to higher-order schemes. For both phases, solution stability is further supported by a Courant number ramp, gradually increasing the value up to its default setting. Following this procedure, a minimum of 50 iterations is performed for each incremental step of the Courant number, with a total iteration limit set to 2250. To mitigate potential numerical issues arising from non-smooth geometries, often generated during the optimisation process, an additional convergence criterion is introduced, easing the previous stop conditions. Specifically, the maximum tolerated imbalance in mass flow rate across the pump boundaries is defined as the one resulting in a relative variation in normalised thrust not exceeding $1 \cdot 10^{-4}$. Table 1 compares between the relevant numerical configurations of single- and two-phase simulations.

Table 1. Summary of the numerical setup for single- and two-phase simulations.

	1-Phase	**2-Phase**
Turbulence model	$\kappa - \omega$ SST	$\kappa - \omega$ SST
Cavitation model	–	Zwart
Momentum discretisation	second order upwind	second order upwind
Turbulence discretisation	second order upwind	second order upwind
Pressure discretisation	second order upwind	PRESTO!
Volume fraction discretisation	second order upwind	QUICK
Free-stream inflow	velocity inlet	velocity inlet
Free-stream outflow	pressure outlet	pressure outlet
Pump inlet	mass flow outlet	target mass flow rate
Pump outlet	mass flow inlet	mass flow inlet
Solid boundaries	no-slip walls	no-slip walls
Courant number	fixed (default)	increasing ramp

On average, a converged simulated case took approximately 12 min on a single-node architecture, equipped with 48 Intel CascadeLake 8260 processors @2.4 GHz.

2.3. Thrust-Drag Bookkeeping

Providing a direct definition of the processed streamtube for axisymmetric propulsors, the aero-engine Thrust-Drag Bookkeeping (TDB) offers a convenient framework for force accounting in ODW investigations. This technique shares several similarities with the method proposed by 24th I.T.T.C. [46], traditionally employed in marine propulsion studies, yet it overcomes specific limitations, making it a more suitable approach for ODW applications, as motivated in Avanzi et al. [21].

Following typical TDB implementations [47], the adopted notation reflects a clear distinction between force components acting on walls located either inside (θ) or outside (ϕ) the thrust domain-an axisymmetric volume bounded by the stagnation and released streamlines (Figure 3). On fluid-dynamic interfaces, gauge forces (FG) are applied based on momentum integration.

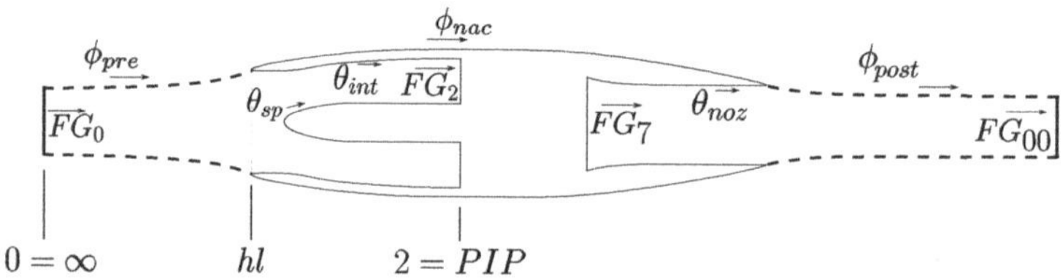

Figure 3. Schematic of ODW Thrust-Drag Bookkeeping (TDB) domain with relevant quantities.

The corresponding expressions at a generic surface are reported below:

$$\phi, \theta = \hat{x} \cdot \iint [(p - p_\infty)\hat{n} + \bar{\tau}_w \cdot \hat{n}] dA \tag{2}$$

$$FG = \hat{x} \cdot \iint \left[(p - p_\infty)\hat{n} + \rho \vec{V} (\vec{V} \cdot \hat{n})\right] dA \tag{3}$$

where p and p_∞ are the field and free-stream pressures, $\hat{n}$ is the surface unit normal, $\bar{\tau}_w$ is the shear stress tensor, ρ is the fluid density and $\vec{V}$ is the velocity field. Then, a reference direction ($\hat{x}$) is considered to project forces. Assuming an axisymmetric flow evolution, thrust can be considered as aligned along the advancing, axial direction, which will be adopted for projection.

Applying TDB definitions to the reference domain in Figure 3, the nacelle drag, D_{nac}, is computed as:

$$D_{nac} = \phi_{pre} + \phi_{nac} + \phi_{post} \tag{4}$$

where the pre-entry (ϕ_{pre}) and post-exit (ϕ_{post}) streamtube forces are required to defined a closed-volume region for the integration. Concerning the former, momentum balance can be conveniently employed to derive a straightforward expression, reading as:

$$\phi_{pre} = FG_2 - FG_0 + \theta_{int} + \theta_{sp} \tag{5}$$

where the free-stream gauge force computation is analytically given as $FG_0 = \rho Q v_\infty$. Differently, the post-exit term can only be computed using Equation (2) along the released streamline.

Then, a non-dimensional drag coefficient, $c_{D_{nac}}$, can be defined upon the PIP cross-sectional area, A_{PIP}, as:

$$c_{D_{nac}} = \frac{2D_{nac}}{\rho v_\infty^2 A_{PIP}} \tag{6}$$

Additionally, computing the force components along all the near-field propulsor's boundaries, the system net force, T, can be obtained as:

$$T = FG_7 - FG_2 - (\theta_{int} + \theta_{sp} + \phi_{nac} + \theta_{noz}) \tag{7}$$

Finally, this parameter can be used to compute the propulsive efficiency as:

$$\eta_{prop} = \frac{T v_\infty}{Q \Delta(p^0)_{pump}} \tag{8}$$

with $\Delta(p^0)_{pump}$ denoting the total pressure jump across the pump boundaries.

2.4. Grid Sensitivity

While the validation of the intake model was already presented in previous works by the authors [20,21], no experimental data are currently available for comparison regarding the external nozzle wall flow distribution. However, the same numerical approach proved reliable for analysing all other components of the configuration. Therefore, this study extends the use of these techniques, assuming their applicability for predicting the fluid dynamic behaviour of the complete arrangement. Specifically, three grid refinements are adopted to assess the dependence of the computational outcomes on the spatial resolution. Coarse, medium and fine meshes account for, respectively: 82k, 184k and 441k cells.

The Grid Convergence Index (GCI) [48] is computed to quantify the discretisation error in estimating computed parameters. The index is evaluated for two relevant flow configurations at both cruise and near-take-off conditions to properly drive the resolution selection according to the tasks expected during the optimisation. A flow configuration is completely defined once three quantities are defined, respectively: the IVR, the highlight diameter based Reynolds number, Re_{hl}, and the cavitation number, σ. These latter are expressed as follows:

$$Re_{hl} = \frac{\rho v_\infty D_{hl}}{\mu} \tag{9}$$

$$\sigma = \frac{2(p_\infty - p_v)}{\rho v_\infty^2} \tag{10}$$

where D_{hl} is the highlight diameter and μ is the dynamic viscosity.

The flow field of the two operating points, obtained with the coarse grid, is reported in Figure 4 using contours of the pressure coefficient, $c_p = 2(p - p_\infty)/(\rho_\infty v_\infty^2)$. Near take-off (Figure 4a), characterised by $Re_{hl} = 2.7 \cdot 10^6$ and $\sigma = 1.64$, the streamtube at the capture section features a convergent shape, consistent with $IVR = 1.11$, indicating an acceleration of the flow entering the inlet opening. Despite the shift of the stagnation point toward the external wall, the major pressure decay occurs just downstream of the highlight lip, leading to the formation of a small, localised vapour bubble before the inlet throat. However, this phenomenon does not noticeably affect the overall performance. Differently, at near-design conditions (Figure 4b), a nominal flow distribution in the proximity of the nacelle leading-edge ensures the absence of cavitation onset. In fact, at these cruise reference flow conditions ($Re_{hl} = 3.4 \cdot 10^6$ and $\sigma = 1.02$), the capture streamtube, aligning with $IVR = 0.85$, decelerates before reaching the highlight section, thus preventing local phase transitions. At the exhaust section, flow evolution is constrained by the straight duct configuration, with the throat area determined according to the assumption of a fixed pump. As a result, under both operating conditions, the pressure field within the nozzle remains mostly bounded within the range of free-stream values.

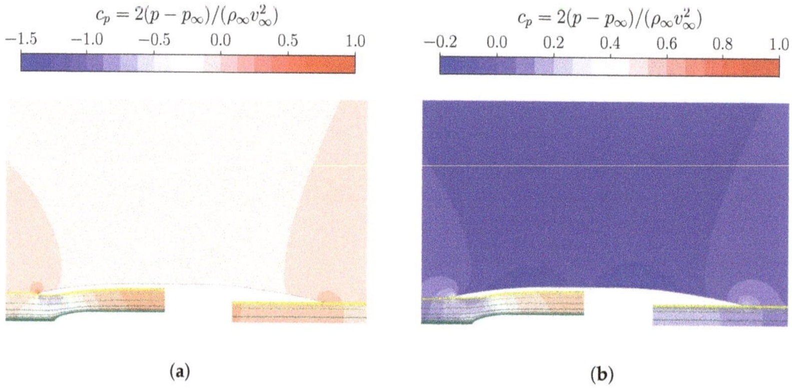

Figure 4. Contours of pressure coefficient, c_p, superimposed with thrust domain streamlines, at both off-design, $IVR = 1.11$, (**a**) and near-design, $IVR = 0.85$, (**b**) operating conditions.

The GCI is evaluated for both the system's drag coefficient at cruise conditions and the Pressure Recovery (PR), $PR = p^0_{PIP}/p^0_\infty$, at off-design conditions, with p^0 denoting the stagnation pressure. Following the procedure described in Celik et al. [48], these parameters are denoted by φ, with subscripts indicating the specific refinement level, ranging from 3 (coarse) to 1 (fine). The results of this analysis are summarised in Table 2.

Table 2. Grid convergence analysis at cruise and off-design operations. The corresponding statistics are the nacelle drag coefficient, $c_{D_{nac}}$, and the pressure recovery, PR.

	$c_{D_{nac}}\vert_{cruise}$	$PR\vert_{off-design}$
φ_3	0.0588	0.9853
φ_2	0.0587	0.9805
φ_1	0.0586	0.9781
φ^{32}_{ext}	0.0581	0.9763
e^{32}_{ext}	0.0113	0.0043
GCI^{32}_{medium}	1.394%	0.533%
φ^{21}_{ext}	0.0578	0.9763
e^{21}_{ext}	0.0137	0.0019
GCI^{21}_{fine}	1.685%	0.232%

The three mesh levels display nearly equivalent performance in predicting propulsive characteristics, with cruise drag and off-design PR GCI indexes stably below 2% and 1%, respectively. Although the coarsest grid would offer faster computations without compromising accuracy, the medium-resolution mesh is retained for the following analyses. This conservative choice accounts for potentially non-optimal geometries generated during the optimisation process, which may be more sensitive to local spatial resolution. Retaining a finer grid ensures a more robust evaluation and allows for the inclusion of a wider range of solutions, enriching the statistical significance of the optimisation outcomes.

3. Results and Discussion

3.1. Propulsor Maps Characterisation

A comprehensive characterisation of the propulsor's operating points is obtained through systematic analyses across a variety of conditions. The aim is to accurately identify critical scenarios to serve as drivers for the multi-point optimisation process. The investigation begins with a single-phase model to map the nominal performance of the propulsor under ambient conditions that prevent cavitation onset.

To trace the characteristic speedlines, the free-stream velocity is held constant, while the mass flow rate boundary condition across the machine is varied in each simulation. Precisely, IVR is adopted as independent variable and the mass flow rate is estimated from Equation (1) according to the advance condition. This procedure is applied to four different intake geometries: the baseline and three optimised designs obtained in a previous study [21] and labelled as *opt1*, *opt2* and *opt3*. These shapes are selected from opposite ends of the Pareto front resulting from a two-objective optimisation conducted in that study, which targeted forebody drag and PR under cruise conditions. All configurations are paired with the same nozzle profile to eliminate variability associated with exhaust dynamics (Figure 5).

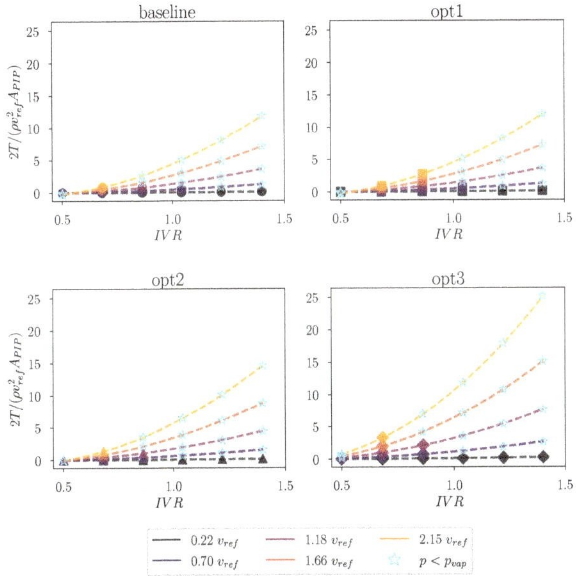

Figure 5. Propulsor's operating maps comparing three optimised intake shapes from Avanzi et al. [21] and the baseline, coupled with the same nozzle geometry. Simulations are performed with single-phase model and the net force, T, is normalised with the baseline cruise reference velocity.

In contrast, the variation in intake geometry is intentionally used to guide the selection of the off-design operating point, capturing the associated effects relevant to the optimisation process. Specifically, the optimised individuals feature progressively increasing highlight areas compared to the baseline. The resulting data are expressed in non-dimensional units, adopting the IVR as a measure of the flow rate and a thrust coefficient computed as $2T/(\rho v_{ref}^2 A_{PIP})$. Here, the reference velocity, v_{ref}, corresponds to the cruise condition of the original design presented in Avanzi et al. [21], which is consistent with $Re_{hl} = 3.4 \times 10^6$. Although multiple advancing conditions are analysed, the same normalisation parameter is adopted across all speedlines. While the thrust coefficient is conventionally defined with respect to the specific free-stream velocity, such a choice in this case would cause all curves to collapse into a single linear trend with respect to IVR, thus preventing a distinction between different speedlines. This approach also enables a clear identification of critical operating points with potential for cavitation onset, denoted with blue star labels when pressures lower than the vapour value, p_{vap}, are detected. Observing the behaviour of the parametric curves for any geometry, it is possible to infer that the net thrust has a monotonic smooth trend with respect to the mass flow rate variation. Data can be fitted with a quadratic law, where the second-degree term coefficient has a positive correlation with the advancing speed. In fact, as this parameter increases, the parabola concavity becomes more emphasised, inducing steeper increment of the non-dimensional thrust with augmenting IVR. However, the curve shape is not only dependent on the free stream conditions, but also on the geometry outline. Specifically, as the highlight area is increased from the baseline to *opt3*, the sensitivity to IVR becomes stronger. Especially for the last individual, this dependency determines over a 60% increment of the last operating point at the highest speed, compared to the same condition in *opt2*.

Considering that the exhaust region exhibits the same dynamics for any configuration, the propulsive performance different characters are driven by the inlet geometry. Specifically, under the same advancing conditions, an equal IVR corresponds to higher mass flow rates processed by wider inlet highlight sections. Consequently, the local flow acceleration becomes stronger, inducing lower wall pressures. Since no cavitation is modelled at this stage, pressure values may reach negative values, with no lower limit restricting the evolution of the variable. The related integration results in a positive contribution to thrust, which motivates the gradual up-shift of the curves as the capture sections enlarges from the baseline to the last optimised individual. Evidently, these geometries exhibit greater susceptibility to cavitation onset. However, this behaviour is closely linked to detailed shape features, as critical pressures tend to develop in localised regions depending on the specific wall curvature. Anyhow, cavitation volumes can potentially evolve externally for low mass flow rate operations, while internal regions may be affected at high IVR. Specifically, the phenomenon is clearly influenced by the advancing condition, according to Equation (10). As a results, low-speed operations are in general safer across wider ranges of processed mass flow rates, while, as the free stream velocity increases, cavitation becomes more likely, with clear dependency on the local wall shape.

To properly investigate the propulsive behaviour of the system, the same simulations are performed with a two-phase numerical model, including Zwart cavitation formulation. Considering the increased complexity of the flow field evolution, the solution becomes more unstable. Thus, in some simulated points, either divergence or incompatibility with the mentioned convergence criteria determine the rejection of the computed solution. This generally occurs when large cavitation volumes form internally, while external bubbles can be solved with fewer computational issues.

Figure 6 reports the PR as a function of the IVR for the same geometries and free stream conditions analysed in the previous discussion. The results confirm a stronger tendency for cavitation onset at higher free-stream velocities. At low advancing speeds, pressure recovery shows a smooth decreasing trend with increasing IVR, consistent with the evolution of the captured streamtube. Specifically, when the stagnation point is located inside the intake, the diffuser operates under higher pressure levels, enhancing recovery at the PIP. As the stagnation point moves outward with increasing flow rates, the diffusion process weakens, leading to reduced PR values. This trend is highly dependent on the free-stream conditions. As the far-field velocity increases, the PR decay becomes more rapid, triggering local phase transitions that cause sudden performance drops. Consequently, as larger capture sections promote local accelerations, the maximum IVR for which convergence is achieved tends to decrease when simulating optimised geometries. The effect of geometry, however, remains strongly localised, preventing the definition of general predictive laws. Conversely, external cavitation has negligible influence on the current performance parameter. Still, increased capture velocities are associated with greater entrance losses, leading to lower PR values at equivalent IVR conditions.

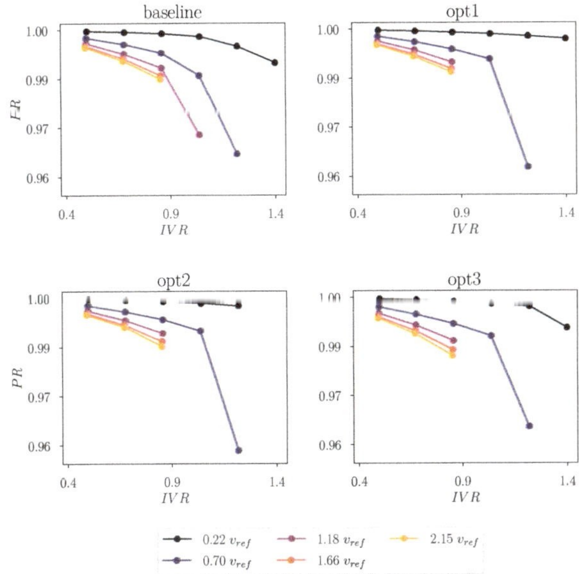

Figure 6. Distributions of the pressure recovery, PR, as a function of the IVR, comparing three optimised geometries from Avanzi et al. [21] and the baseline. Data obtained from two-phase simulations at several advancing velocities.

Internal lip cavitation is further examined through contours of vapour volume fractions (Figure 7). The analysis considers operating conditions at $v_\infty = 0.70\ v_{ref}$ with $IVR = 1.22$, comparing among three geometries: the baseline (Figure 7a), *opt1* (Figure 7b) and *opt3* (Figure 7c). In fact, although featuring a smaller highlight area than the last geometry, *opt2* local cavitation evolution prevents the satisfaction of the continuity criterion across the machine.

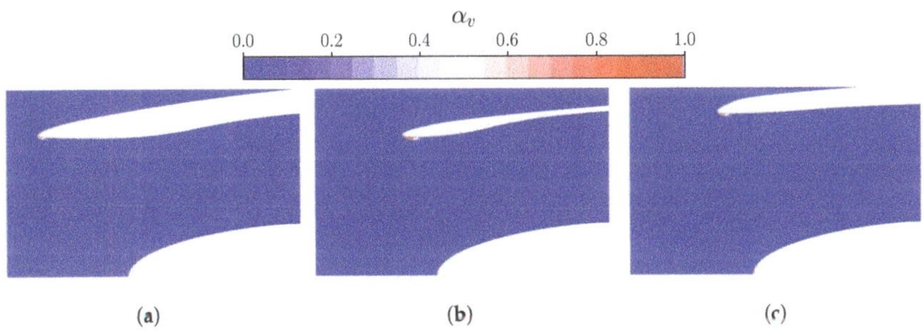

Figure 7. Vapour volume fraction distributions for baseline (**a**), *opt1* (**b**) and *opt3* (**c**). Internal cavitation is visualised at $IVR = 1.22$ and $v_\infty = 0.70\ v_{ref}$.

For the reference geometry, vapour bubbles rapidly form near the sharp leading-edge and extend downstream over a limited distance, without reaching the inlet throat. While a rounded edge can mitigate abrupt cavitation development-as seen in the *opt1* flow field-the shorter lip length in this case facilitates phase transition. This is because the minimum duct section is located closer to the stagnation point, sustaining local acceleration further downstream. As a result, thicker and longer vapour regions develop, extending beyond the throat and directly reducing pressure recovery compared to the previous configuration. In contrast, the flow field around *opt3* suggests that a blunter lip slightly reduces the extent of the vapour region. This geometry partially limits the acceleration through the channel, leading to lower total pressure losses at the PIP.

As the mass flow rate increases, the vapour bubble enlarges and extends downstream into the diffuser. This structure introduces additional obstruction and distortion to the incoming flow. As a result, the reduced passage area near the throat further accelerates the flow, triggering a non-linear amplification of the phase transition that rapidly compromises solution stability. For this reason, computations tend to diverge earlier, leading to fewer valid simulations.

At the same free-stream conditions, the system's resistance is evaluated by analysing the distribution of the nacelle drag coefficient as a function of the IVR (Figure 8). As previously observed, lower free stream velocities mitigate cavitation evolution. Consequently, from speedlines analysis it follows that the corresponding trends are clearly non-monotonic. In fact, the curves generally agree in recognising a minimum around $IVR = 1$, which would correspond to the optimal condition predicted by the potential flow theory. Again, the smooth behaviour of each speedline is corrupted as cavitation occurs, which is typical of flow rate conditions different from nominal. At the early stages of internal cavitation, when $IVR > 1$, lip phase transition locally reduces the drag component, promoting the overall performance. However, this evolution is only beneficial at low velocities ($IVR = 1.22$, $v = 0.70v_{ref}$) where the vapour extension is limited. As cavitation covers large portions, potentially extending downstream inside the diffuser, the entering streamtube is obstructed, inducing instability and feeding distortions and local separations that compromise performance. Differently, at low IVR, convergence is generally reached. External cavitation significantly affects nacelle resistance, causing abrupt increases in the drag coefficient, with a stronger dependency observed as the free-stream velocity increases. This trend is further supported by the observation that external cavitation has a lesser impact on solution stability, allowing valid simulations even when vapour regions become substantially more extended compared to those formed within internal structures.

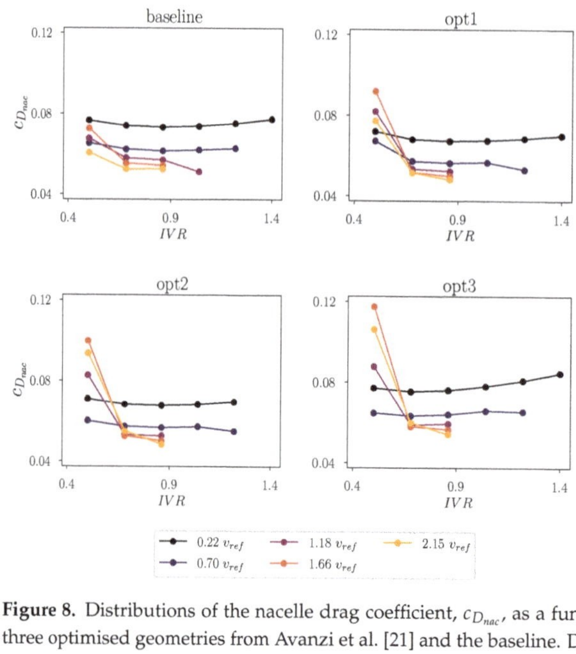

Figure 8. Distributions of the nacelle drag coefficient, $c_{D_{nac}}$, as a function of the IVR, comparing three optimised geometries from Avanzi et al. [21] and the baseline. Data obtained from two-phase simulations at several advancing velocities.

3.2. Design Space Definition and Setup

To perform a systematic analysis using a DOE approach, an automatic procedure is used. Specifically, once a database of geometrical variants is generated, the corresponding individuals need to be discretised and simulated using CFD before post-processing. To easily manipulate the geometry, reducing the number of Decision Variables (DV), the original shape is first parameterised using B-splines. Each control point is assigned a specific Degree of Freedom (DOF), which determines its allowed displacement in the x and/or y direction from its original position in the reference parameterisation (Figure 9). Bounding control points, such as the axial and radial coordinates of the highlight radius and the axial coordinate of the nozzle Trailing Edge (TE), are directly modified. Differently, internal control points are first linearly scaled to match the updated global extension. Subsequently, they are further adjusted based on locally defined ranges, specifically determined for each control point. These local bounds are illustrated in Figure 9a, either as lines or rectangles depending on whether the number of DOFs is 1 or 2, respectively. Concerning the centerbody (Figure 9b), the geometric variants only account for a rigid axial translation, exerted by applying the same x shift to all control points, while, as aforementioned, the pump sections are left unchanged. Based on this parameterisation, the resulting number of DV adopted for the optimisation is 21. An automatic grid generation process is adopted for each individual, which is characterised by a precise and unique set of DV. The mesh sizing is controlled based on the baseline medium refinement, adjusting local nodes distributions according to the specific B-spline extension. Thus, each case features its peculiar cells count, with an overall range between 166k and 210k. For the multi-point investigation, two operating conditions are considered: a near-cruise configuration with $Re_{hl} = 3.4 \cdot 10^6$ and $\sigma = 1.02$, and an off-design scenario with $Re_{hl} = 2.7 \cdot 10^6$ and $\sigma = 1.64$. Considering the baseline geometry, these points, from CFD simulations, correspond to $IVR = 0.85$ and 1.11, respectively. The same conditions are analysed, considering the potential integration between the pump and the propulsor (Figure 10). The ODW maps are estimated using a meanline approach for different advancing velocities and plotted over the pump characteristic curves, which are propagated from the non-dimensional law for different rotor velocities, n. The machine flow rate, head ($H = \Delta(p^0)_{pump}/(\rho g)$) and rotational regimes are normalised with the corresponding design values, Q_d, H_d and n_d. The selected operating points underscore that, under the chosen off design conditions, the IVR constrains the performance required from the pump, specifically indicating operations at higher flow rates and rotor speeds. Here, the discrepancy in the reported IVR value depends on the different accuracy between the meanline method and CFD. Table 3 summarises the flow configurations corresponding to the two operating points selected for the optimisation, where simulations are performed according to the solution strategy outlined in Section 2.2.

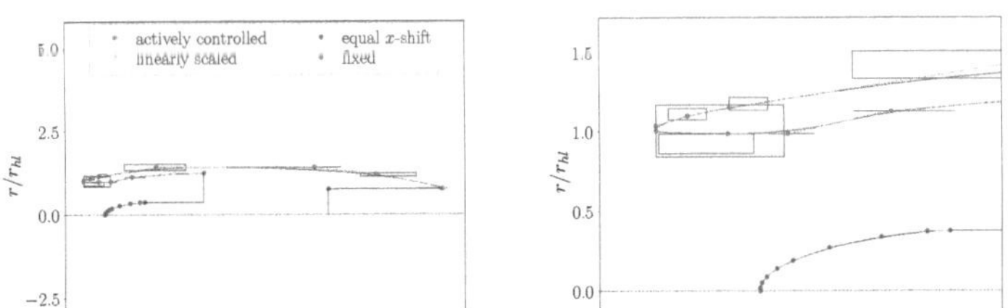

Figure 9. Baseline geometry B-spline parameterisation, with decision variables variation ranges superimposed (**a**) and magnified view near the inlet section (**b**). Control points specific Degrees Of Freedom (DOFs) are rendered using colours and labelled in the legend.

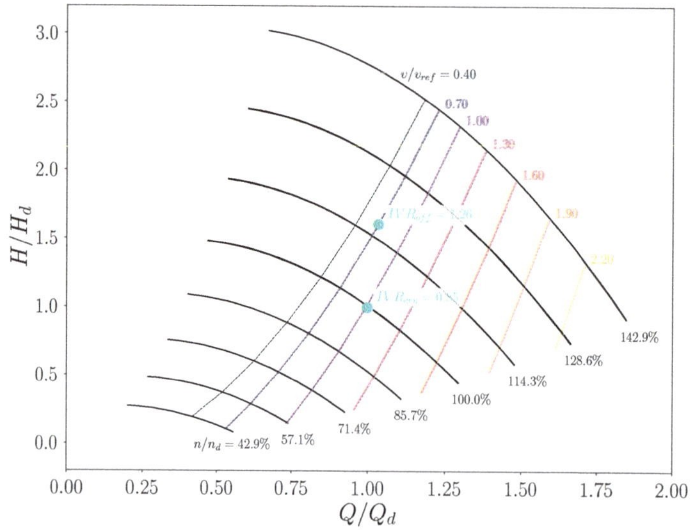

Figure 10. Reference pump characteristic map, normalised by design values, for different rotational regimes and contours of the propulsor's map for different free stream velocities, derived from meanline method. Selected operating points are reported.

Table 3. Summary of the flow configurations of the two operating points adopted for the optimisation.

	Cruise	Off-Design
Re_{hl}	$3.4 \cdot 10^6$	$2.7 \cdot 10^6$
σ	1.02	1.64
IVR	0.85	1.11

The DOE database comprises 512 individuals, generated using a Sobol sequence [49]. The design space spanned by the resulting geometric variants is illustrated in Figure 11, where the exploration bounds of key global dimensions are shown relative to their baseline values. The chart also depicts the distribution of converged individuals according to the acceptable simulation outcome. It is clear that, while a large number of valid solutions are retained, the proportion of non-converged cases remains non-negligible. A potential correlation between simulation invalidity and geometric configuration emerges when considering the intake-specific cross-sectional dimensions. Two key parameters are examined: the Contraction Ratio, $CR = A_{th}/A_{hl}$—where A_{th} is the throat area—and the highlight radius. Although the overall sampling is symmetric about the baseline, the invalid cases appear predominantly clustered in the lower-left region of this subspace. This is attributed to the boundary condition setup: with a fixed mass flow rate, reducing both CR and highlight radius increases the IVR, potentially leading to critical lip accelerations. As a result, convergence is often not achieved at cruise conditions, and even less so under off-design scenarios. Conversely, as these two parameters increase, the number of valid individuals rises. Initially, these include cases converging only at cruise, while further enlargement allows convergence under off-design conditions as well. At the upper-right corner of the parameter space, some individuals exhibit convergence exclusively at off-design. This outcome is associated with excessively low cruise IVR values, which destabilize the flow along the external wall—often due to extensive cavitation or flow separation—thus promoting off-nominal operating regimes. Regarding axial-extension parameters, such as the nacelle length (l_{nac}) and the nozzle length (l_{noz}), the design space is largely populated by compact nacelle configurations. While the database includes geometries with longer nozzles, most individuals correspond to shorter overall system lengths, with relatively few solutions representing longer configurations.

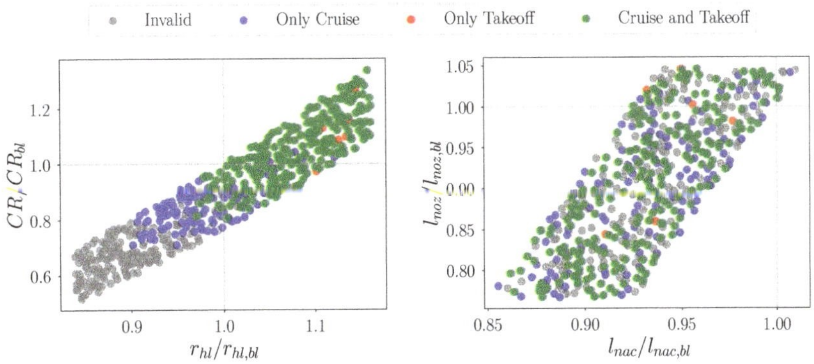

Figure 11. Distribution of the decision variables design space, visualised through relevant global dimensions, normalised by the corresponding baseline value. Dots are colourised based on the convergence state of both cruise and off-design simulations.

3.3. Optimal Geometries for Objectives Optimisation

The results of the simulated geometries are analysed to detect potential correlations between propulsive statistics and relevant shape parameters at both in- and off-design operations. Specifically, the recovery sensitivity is initially investigated considering intake relevant dimensions (Figure 12). These include the highlight radius, the contraction ratio and the equivalent diffuser angle, α_{diff}, computed from the axial (x) and radial (r) coordinates of the PIP and the throat sections, according to the following expression:

$$\alpha_{diff} = 2\arctan\frac{r_{PIP} - r_{th}}{x_{PIP} - x_{th}} \quad (11)$$

At design operations (Figure 12a), the propulsive statistics exhibits an evident correlation with both r_{hl} and the IVR. In fact, under the same flow configuration, increasing the geometric parameter results in lower streamtube capture ratios. This shape variation shows to improve the recovery capability, an evidence which is further confirmed by the reduction of the dots dispersion and the dissipation of outliers. Enhanced PR can also be achieved with larger contraction ratios. However, this approach does not show a clear correlation with IVR, apart from a tendency for lower IVR values to cluster toward the right side of the chart. Notably, the individuals distribute on a narrow band, clearly indicating a non-linear trend with reduced steepness at higher CR values. Conversely, smaller diffuser angles are required to increase the recovery capability. The overall trend again displays a non-linear behaviour, with increased dispersion as the geometric parameter grows—an effect that approximately corresponds to higher capture ratios.

Performing equivalent statistics at off-design conditions, similar considerations can be drawn (Figure 12b). In fact, the PR preserves the same dependencies on the geometrical parameters. However, it is interesting to notice that, in this case, all charts depict distinct correlations, made clearer by considerably narrower dispersion bands. On the other hand, a greater number of outliers arises. Even the baseline solution is evidently outside the clustering of the database individuals, especially when CR and α_{diff} sensitivities are considered. This aspect highlights that most individuals' PR values lie above the reference geometry's performance. Moreover, the trend is flatter compared to the previous analysis. This evidence suggests a wide room for optimisation at non-nominal operations, with potential to improve the parameter by at least 2%. This represents a clear advancement beyond the cruise design improvements reported in previous studies [21]. This combined multi-point investigation reveals that the two objectives are mutually consistent, indicating that a geometry improving cruise operations can also optimise off-design performance.

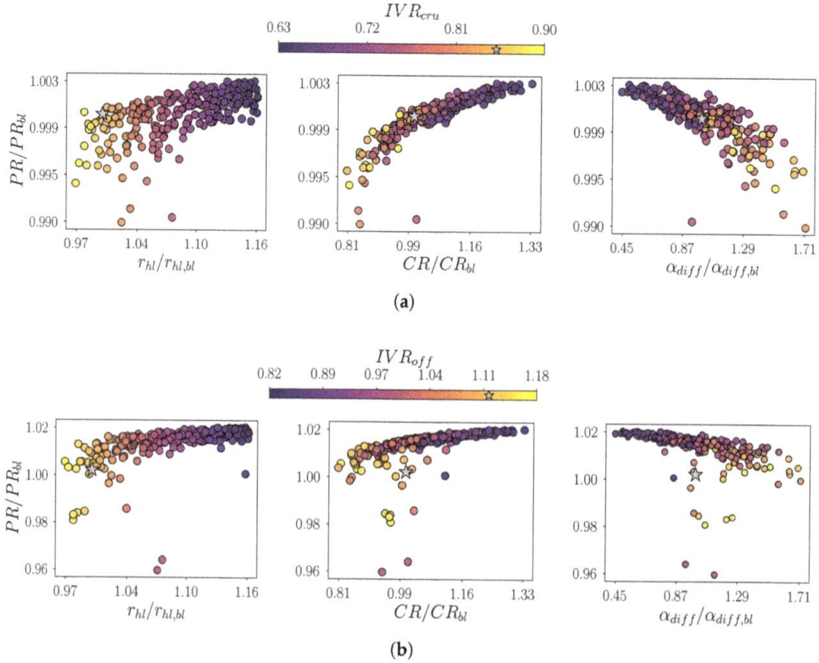

Figure 12. Sensitivity of the pressure recovery to relevant intake geometrical parameters at both in- (**a**) and off-design (**b**) operating conditions.

To trace the impact of the overall geometry variation, an equivalent analysis is performed on the nacelle drag, D_{nac} (Figure 13). In this case, the chosen geometrical parameters are: the intake length, $l_{int} = x_{PIP} - x_{hl}$, the maximum nacelle radius, r_{max}, and the trailing edge cowl angle, β_{TE}, which can be derived directly as the slope of the segment connecting the ending control points of the external nozzle B-spline.

For in-design operations (Figure 13a), the statistic's distribution suggests that shorter intakes generally promote drag reduction. While a significant number of outliers exceed the baseline drag, the majority of individuals cluster within a denser region. This latter exhibits a quasi-linear trend, indicating that reducing axial extension generally leads to design improvement. A comparable pattern emerges when analysing the influence of the maximum nacelle cross-sectional dimension. Although the variations are symmetrically distributed around the reference value, the results reveal a consistent reduction in nacelle drag with decreasing diameter—attributable to an overall thinner body. In contrast, the sensitivity to the boat tail angle presents a more scattered distribution. Most geometries group near the baseline, showing both improved and degraded performance. Although the geometric features of the shape variants complicate establishing a clear correlation, the wider spread across the β_{TE} range reveals a discernible trend. Steeper wall gradients generally correspond to higher drag on the rear nacelle section, thereby degrading overall performance. Finally, no meaningful correlations are observed between IVR and any of the three geometric parameters in the charts.

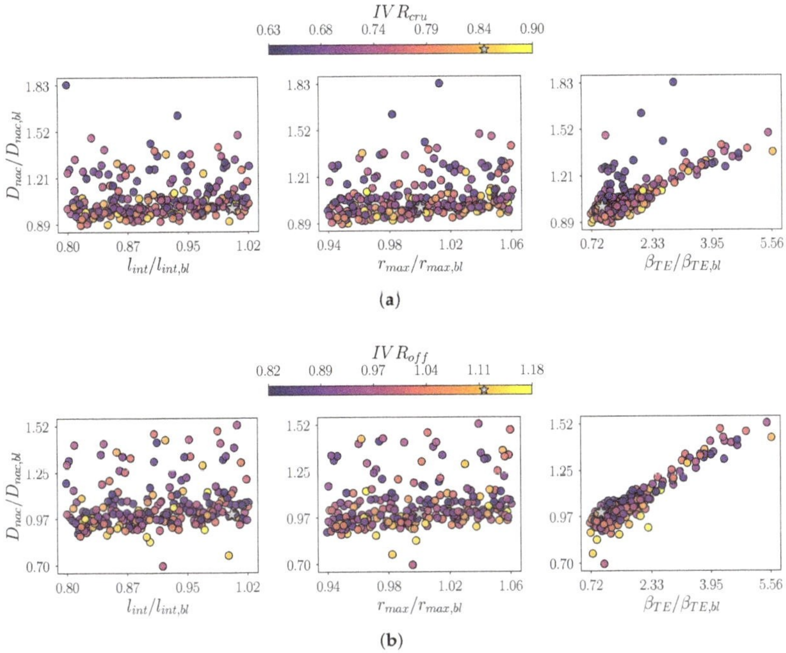

Figure 13. Sensitivity of the nacelle drag to relevant shroud geometrical parameters at both in- (**a**) and off-design (**b**) operating conditions.

The analysis of statistics related to off-design operations (Figure 13b) confirms previous observations regarding PR. In fact, the sensitivities to the three parameters exhibit trends similar to those observed for cruise performance. However, most of the individuals cluster around values that have lower improvements compared to nominal conditions. Interestingly, in this case all the charts show the presence of few outliers providing significantly better performance, while the ones featuring higher drag are considerably more. Unlike pressure recovery, potential improvements at off-design operations are limited. However, the consistency with cruise design enhancements allows reducing the analysis effort by focusing on a single operating point for each performance metric.

The previous analysis showed that both the PR and the nacelle drag exhibit the same sensitivity to the relevant geometrical parameters at the chosen operating points. According to this consideration, the problem is further restricted and the investigation is progressed using a single objective for each advancing regime. Specifically, the PR showed to be a fundamental statistic to be considered during off-design operations. Additionally, its tendency to be easily optimised under this operating regime makes it a convenient choice. On the other hand, maximising the net force during long-range operations is recommended to improve propulsive efficiency, which is also supported by the wider potential for drag minimisation at cruise conditions. Plotting the two objectives function space confirms their conflicting nature as a Pareto front emerges on the upper right region (Figure 14). To preserve the information regarding the excluded statistics, only a selected set of individuals is reported. Specifically, only the geometries featuring cruise $PR/PR_{bl} \geq 1$ are retained, while the off-design net force is adopted to colourise dots.

The chart indicates that almost the entire set of individuals improves off-design pressure recovery. Concerning thrust, many geometries exhibit worse performance, even though the majority groups at enhanced operations up to +2%. Analysing the colour trend, its smooth distribution confirms that maximising cruise performance is also beneficial for off-design conditions, ensuring a similar range of potential improvements.

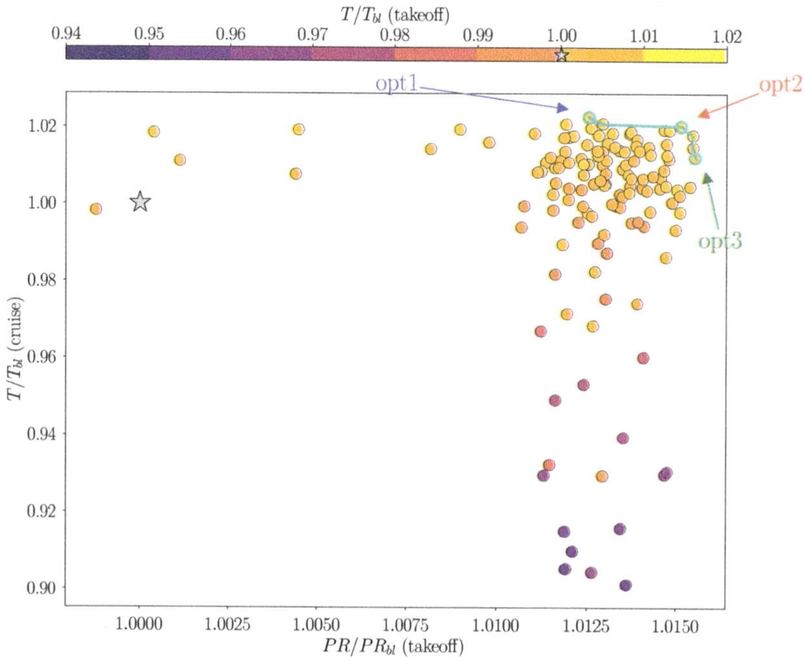

Figure 14. Pareto front (green line) emerging from the objectives function space, determined by the pressure recovery at off-design and net force at cruise operations.

Three individuals are then selected for additional investigation, to properly characterise how geometrical properties affect the propulsive objective functions. They are chosen from different locations on the Pareto front, which enables to focus on the proper features of optimal individuals. The first and the last individuals, respectively labelled as *opt1* and *opt3*, favour opposite objectives. Specifically, the former has the best drag performance, while the latter provides the highest PR increment. A third geometry, named *opt2*, is chosen at mild conditions to characterise trade-off solutions. The three geometries are compared with the baseline in Figure 15. Maximum net force solution is characterised by a short intake configuration (−16%), where the lip extension is minimised and the spinner tip is aligned with the throat section. Compared to the baseline, the highlight section is significantly enlarged by 23% with blunter leading-edge. The internal diffuser curvature results smoother, while the nozzle extension appears considerably reduced, leading to an overall length decrease by 12%. On the other hand, *opt3* features a slightly lower intake length reduction, even though the capture section is further increased (+34%). This results in a overall flat profile of the internal nacelle wall, with a considerably larger lip axial extension. In this case, the sharp profile at the leading-edge is preserved, while the spinner is significantly retracted and separated from the throat section. Concerning the trade-off geometry, *opt2*, its specific properties partially combines the others' characters. In fact, while the nozzle length and capture sections feature mild sizes, the intake length is evidently the smallest among the individuals (−20%). The spinner location, instead, is almost the same as *opt1*.

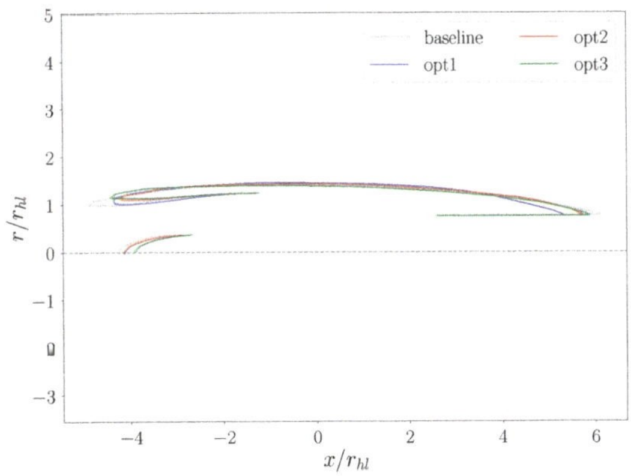

Figure 15. Geometrical comparison between optimised individuals and baseline. The corresponding labels refer to Figure 14.

To analyse the flow field distribution locally, the wall pressure coefficient is extracted along the nacelle's and spinner's walls. During design operations (Figure 16), the flow acceleration on the external surface (Figure 16a) induces similar pressure coefficients values. Despite considerably exceeding the baseline negative coefficient, they do not reach vapour condition: in fact, according to Equation (10), at cavitation onset $\sigma = -c_p$. The evolution then rapidly re-aligns with baseline pressure field, depicting local overshoots, particularly pronounced in the case of *opt2*. It should be noted that the use of random sampling for optimisation may introduce irregular local profile curvatures, which could be avoided through more refined, direct approaches.

Major differences can be detected along the internal surface. Specifically, *opt1* exhibits a generally lower curve, gradually matching the other solutions toward the PIP. Especially in the contraction region, this individual exploits higher local acceleration to reduce the resistance force pressure component. This evolution is particularly effective on the spinner walls as well (Figure 16b). In fact, the pressure distribution is generally lower along the entire profile compared to any other geometry. This enables a consistent reduction of the drag force related to the spinner, favouring the overall net force maximisation.

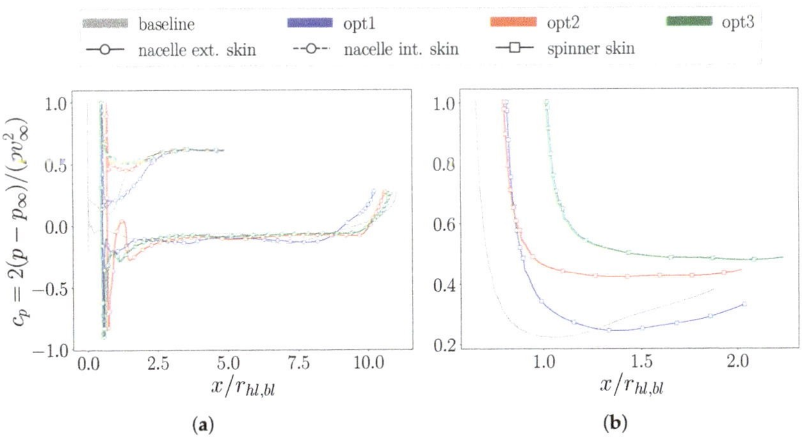

Figure 16. Pressure coefficient distribution along the nacelle's (**a**) and the spinner (**b**) walls, comparing individuals and baseline operations at cruise.

The evolution at off-design operations is considerably different (Figure 17). Specifically, as the the stagnation point shifts outward, local accelerations on the external walls result mitigated (Figure 17a). Conversely, stronger pressure drops occur on the internal surfaces. As a results, the risk of cavitation onset augments in the internal duct, critically inducing the formation of small local vapour bubbles as in the case of baseline geometry. While any other shape prevents this occurrence, the pressure variations are consistent, leading to higher depressions inside the diffuser compared to the outer shroud. The main consequence is that the inlet diffusion process is compromised by highly distorted flows which introduces additional losses, thereby affecting the pressure recovery. In this regard, the advantage of *opt3* is that the internal flow is only marginally influenced compared to cruise conditions. In fact, despite exhibiting lower values, the coefficient curve depicts a similar smooth and flat behaviour. A similar consideration applies to the evolution along the spinner walls (Figure 17b). As a result, by controlling flow perturbations throughout the internal channel, this geometry ensures high recovery capability when the propulsor processes a non-optimal capture streamtube. However, this comes at the cost of compromised propulsive efficiency.

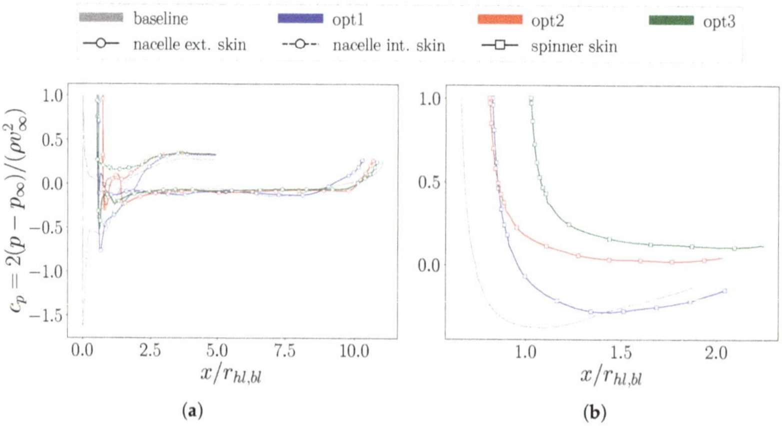

Figure 17. Pressure coefficient distribution along the nacelle's (**a**) and the spinner (**b**) walls, comparing individuals and baseline operations at off-design.

4. Conclusions

This study investigates multi-point design aspects required for optimal shaping of Outboard Dynamic-inlet Waterjets (ODW), a novel marine propulsion unit that can operate isolated from the ship. Aiming to guide the nacelle definition through insightful correlations between fundamental geometric features and key propulsive performance statistics, a systematic analysis is conducted to balance conflicting operational performance. a systematic analysis is conducted to balance conflicting operational requirements. In particular, two parameters demand focused attention for achieving efficient mission envelopes: minimisation of system drag to enhance net thrust during cruise operations, and maximisation of inlet pressure recovery to ensure sufficient cavitation margin for the pump under off-design conditions. To address these objectives, an optimisation study is performed, adopting a 2D axisymmetric, two-phase Computational Fluid Dynamics (CFD) model to simulate the flow field past the propulsor. Thus, a database of geometric variants of a baseline outline is sampled using random Design Of Experiments (DOE). The corresponding CFD solutions are analysed to extract relevant propulsive metrics using the aero-engine Thrust-Drag Bookkeeping (TDB).

Initially, the propulsor's characteristic maps are traced through systematic simulations, varying both free stream conditions and processed mass flow rates. This starting phase underscores that operations at capture ratios higher than nominal may result critical due to the onset of internal lip cavitation, obstructing and distorting the entering streamtube. Based on this characterisation, two relevant operating points are selected-one at cruise and one at take-off-to investigate several geometries and demonstrate how optimal shapes should be designed to ensure safe operations in both scenarios. The resulting database is systematically analysed, showing that the geometrical parameters affect propulsive metrics with consistent outcomes between the two regimes. Thus, the study focuses on two different objectives: off-design Pressure Recovery (PR) and cruise nacelle drag. The corresponding distribution confirms a conflicting trend, showing the emergence of a Pareto front. To characterise the specific impact of geometric features on propulsive performance, three distinct optimised individuals are selected from this region for an in-depth analysis of the corresponding local flow field. The analysis of the intake walls' pressure coefficient distribution indicates that the pressure field within the inlet section plays a key role, directly correlating with the location of the stagnation point. This flow evolution directly influences propulsive performance, showing that shorter nacelles and blunter lips can improve the nominal net force by 2%, while, compromising the propulsive efficiency, wider inlet openings can improve off-design PR over 1.5%.

Although the present investigation introduces novel details on the optimal geometric principles necessary to satisfy the requirements at multiple operating conditions, it is limited by the sampling approach, allowing no direct control on the walls' curves shape. Thus, future studies aim to further advance these outcomes, approaching the optimisation with finer techniques, enabling to control splines sections with higher local accuracy, for improved high-performance geometries.

Author Contributions: Conceptualization, F.A. and A.M.; methodology, F.A. and A.M.; software, F.A.; validation, F.A.; formal analysis, F.A.; investigation, F.A.; data curation, F.A. and A.M.; writing—original draft preparation, F.A., F.D.V. and A.M.; writing—review and editing, F.D.V. and A.M.; visualization, F.A.; supervision, A.M. and F.D.V. All authors have read and agreed to the published version of the manuscript.

Funding: This research received no external funding.

Data Availability Statement: The original contributions presented in this study are included in the article. Further inquiries can be directed to the corresponding author.

Acknowledgments: We acknowledge the CINECA award under the ISCRA initiative, for the availability of high performance computing resources and support. We acknowledge Christian Tempestin for his collaboration in the study as a Master Student.

Conflicts of Interest: The authors declare no conflicts of interest.

Abbreviations

The following abbreviations are used in this manuscript:

CFD	Computational Fluid Dynamics
DOE	Design Of Experiments
DOF	Degrees Of Freedom
DV	Decision Variables
GCI	Grid Convergence Index
IVR	Inlet Velocity Ratio

NPSH	Net Positive Suction Head
ODW	Outboard Dynamic-inlet Waterjet
PIP	Pump Interface Plane
PR	Pressure Recovery
RANS	Reynolds-Averaged Navier Stokes
SST	Shear Stress Transport
TDB	Thrust-Drag Bookkeeping
TE	Trailing Edge

References

1. Allison, J.L. Marine Waterjet Propulsion. *SNAME Trans.* **1993**, *101*, 275–335.
2. Park, W.G.; Yun, H.S.; Chun, H.H.; Kim, M.C. Numerical flow simulation of flush type intake duct of waterjet. *Ocean Eng.* **2005**, *32*, 2107–2120. [CrossRef]
3. Levy, J. The Design of Water-Jet Propulsion Systems for Hydrofoil Craft. *Mar. Technol. SNAME News* **1965**, *2*, 15–25. [CrossRef]
4. Wislicenus, G.F. *Hydrodynamic Design Principles of Pumps and Ducting for Waterjet Propulsion*; Technical Report Report 3990; David W. Taylor Naval Ship Research and Development Center: Bethesda, MD, USA, 1973. Available online: https://apps.dtic.mil/sti/pdfs/AD0775620.pdf (accessed on 28 May 2025).
5. Huang, R.; Dai, Y.; Luo, X.; Wang, Y.; Huang, C. Multi-objective optimization of the flush-type intake duct for a waterjet propulsion system. *Ocean Eng.* **2019**, *187*, 106172. [CrossRef]
6. Chen, L.; Zhou, Y.Y.; Huang, C.L.; Wang, Z.L.; Dai, R. Parametric optimization of vortex generator configuration for flow control in an intake duct for waterjet propulsion. *Ocean Eng.* **2023**, *281*, 114908. [CrossRef]
7. Guo, J.; Zhang, Y.; Chen, Z.; Feng, Y. CFD-based multi-objective optimization of a waterjet-propelled trimaran. *Ocean Eng.* **2020**, *195*, 106755. [CrossRef]
8. Huang, R.; Luo, X.; Ji, B.; Wang, P.; Yu, A.; Zhai, Z.; Zhou, J. Multi-objective optimization of a mixed-flow pump impeller using modified NSGA-II algorithm. *Sci. China Technol. Sci.* **2024**, *58*, 2122–2130. [CrossRef]
9. Long, Y.; Tian, C.; Zhu, R.; Yin, J.; Wang, D. Optimization of water-jet pump based on the coupling of multi-parameter and multi-objective genetic algorithms. *J. Braz. Soc. Mech. Sci. Eng.* **2024**, *46*, 719. [CrossRef]
10. Gaggero, S.; Mastrandrea, F. Improving Waterjet Performance Through Simulation-Based Optimization. *J. Mar. Sci. Eng.* **2025**, *13*, 758. [CrossRef]
11. Huang, X.; Shi, S.; Su, Z.; Tang, W.; Hua, H. Reducing underwater radiated noise of a SUBOFF model propelled by a pump-jet without tip clearance: Numerical simulation. *Ocean Eng.* **2022**, *243*, 110277. [CrossRef]
12. Saghafi, M.; Lavimi, R. Optimal design of nose and tail of an autonomous underwater vehicle hull to reduce drag force using numerical simulation. *Proc. Inst. Mech. Eng. Part M J. Eng. Marit. Environ.* **2020**, *234*, 76–88. [CrossRef]
13. Jiang, J.W.; Huang, W.X. Hydrodynamic design of an advanced submerged propulsion. *Proc. Inst. Mech. Eng. Part C* **2019**, *233*, 6367–6382. [CrossRef]
14. Jiang, J.; Ding, J.; Lyu, N.; Luo, H.; Li, L. Control volume determination for submerged waterjet system in self-propulsion. *Ocean Eng.* **2022**, *265*, 112594. [CrossRef]
15. Wu, C.; Lu, Y.; Liu, S.; Li, Z.; Gu, Z.; Shao, W.; Li, C. Research on Optimization Design of Fully Parameterized Pump-Jet Propulsion. *J. Mar. Sci. Eng.* **2022**, *10*, 766. [CrossRef]
16. Gaggero, S.; Martinelli, M. Design and analysis of pumpjet propulsors using CFD-based optimization. *Ocean Eng.* **2023**, *277*, 114304. [CrossRef]
17. Lu, Y.; Li, C.; Gu, Z.; Wu, C.; Shao, S. Research on multi-objective optimisation design of pump-jet propulsion. *Ocean Eng.* **2025**, *322*, 120086. [CrossRef]
18. Benini, E.; Gobbo, W. Propulsion Device with Outboard Waterjet for Marine Vehicles. European Patent EP3604117A1, 5 February 2020.
19. Avanzi, F.; Benini, E.; Ruaro, F.; Gobbo, W. Design Space Exploration for a Ram Inlet Waterjet for High-Speed Propulsion. In Proceedings of the International Mechanical Engineering Congress and Exposition (IMECE) 2020, Virtual, 16–19 November 2020. [CrossRef]
20. Avanzi, F.; De Vanna, F.; Magrini, A.; Benini, E. Numerical characterisation of Outboard Dynamic-inlet Waterjets: Propulsive aspects of intake/pump integration. *Acta Mech.* **2024**. [CrossRef]
21. Avanzi, F.; De Vanna, F.; Magrini, A.; Benini, E. Two-objective optimisation of an intake diffuser for outboard dynamic-inlet waterjets. *Ocean Eng.* **2025**, *330*, 121187. [CrossRef]

22. Sobolewski, A.D. *Hydrodynamic Performance of the Model of a Variable Area Waterjet Inlet Designed for a 200 Ton, 100 Knot Hydrofoil Ship*; Technical Report SPD-735-O1; David W. Taylor Naval Ship Research and Development Center: Bethesda, MD, USA, 1977. Available online: https://apps.dtic.mil/sti/pdfs/ADA038590.pdf (accessed on 28 May 2025).
23. Lafrance, P. *A Review of Resistance Prediction and Design Methods for Strut-Pod Inlets*; Technical Report SPD-789-O1; David W. Taylor Naval Ship Research and Development Center: Bethesda, MD, USA, 1978. Available online: https://apps.dtic.mil/sti/pdfs/ADA052710.pdf (accessed on 28 May 2025).
24. Michael, T.J.; Schroeder, S.D.; Becnel, A.J. *Design of the ONR AxWJ-2 Axial Flow Water Jet Pump*; Technical Report NSWCCD-50-TR-2008/066; Naval Surface Warfare Center Carderock Division: Bethesda, MD, USA, 2008. Available online: https://apps.dtic.mil/sti/pdfs/ADA489739.pdf (accessed on 28 May 2025).
25. Zhai, S.; Jin, S.; Chen, J.; Liu, Z.; Song, X. CFD-based multi-objective optimization of the duct for a rim-driven thruster. *Ocean Eng.* **2022**, *264*, 112467. [CrossRef]
26. Chesnakas, C.J.; Donnelly, M.J.; Pfitsch, D.W.; Becnel, A.J.; Schroeder, S.D. *Performance Evaluation of the ONR Axial Waterjet 2 (AxWJ-2)*; Technical Report NSWCCD-50-TR-2009/089; Naval Surface Warfare Center Carderock Division: Bethesda, MD, USA, 2009. Available online: https://apps.dtic.mil/sti/pdfs/ADA516369.pdf (accessed on 28 May 2025).
27. Tan, D.; Li, Y.; Wilkes, I.; Vagnoni, E.; Miorini, R.L.; Katz, J. Experimental investigation of the role of large scale cavitating vortical structures in performance breakdown of an axial waterjet pump. *J. Fluids Eng.* **2015**, *137*, 111301. [CrossRef]
28. Cao, P.; Wang, Y.; Kang, C.; Li, G.; Zhang, X. Investigation of the role of non-uniform suction flow in the performance of water-jet pump. *Ocean Eng.* **2017**, *140*, 258–269. [CrossRef]
29. Suryanarayana, C.; Satyanarayana, B.; Ramji, K.; Saiju, A. Experimental evaluation of pumpjet propulsor for an axisymmetric body in wind tunnel. *Int. J. Nav. Archit. Ocean Eng.* **2010**, *2*, 24–33. [CrossRef]
30. Suryanarayana, C.; Satyanarayana, B.; Ramji, K. Performance evaluation of an underwater body and pumpjet by model testing in cavitation tunnel. *Int. J. Nav. Archit. Ocean Eng.* **2010**, *2*, 57–67. [CrossRef]
31. Suryanarayana; Satyanarayana, B.; Ramji, K.; Nageswara, M.R. Cavitation studies on axi-symmetric underwater body with pumpjet propulsor in cavitation tunnel. *Int. J. Nav. Archit. Ocean Eng.* **2010**, *2*, 185–194. [CrossRef]
32. Shi, S.; Tang, W.; Huang, X.; Dong, X.; Hua, H. Experimental and numerical investigations on the flow-induced vibration and acoustic radiation of a pump-jet propulsor model in a water tunnel. *Ocean Eng.* **2022**, *258*, 111736. [CrossRef]
33. Liu, J.; Liu, S.; Wu, Y.; Jiao, L.; Wang, L.; Sun, Y. Numerical investigation of the hump characteristic of a pump–turbine based on an improved cavitation model. *Comput. Fluids* **2012**, *68*, 105–111. [CrossRef]
34. Guo, Q.; Zhou, L.; Wang, Z.; Liu, M.; Cheng, H. Numerical simulation for the tip leakage vortex cavitation. *Ocean Eng.* **2018**, *151*, 71–81. [CrossRef]
35. Guo, Q.; Huang, X.; Qiu, B. Numerical investigation of the blade tip leakage vortex cavitation in a waterjet pump. *Ocean Eng.* **2019**, *187*, 106170. [CrossRef]
36. Avanzi, F.; Baù, A.; De Vanna, F.; Benini, E. Numerical Assessment of a Two-Phase Model for Propulsive Pump Performance Prediction. *Energies* **2023**, *16*, 6592. [CrossRef]
37. Zhao, X.; Liu, T.; Huang, B.; Wang, G. Combined experimental and numerical analysis of cavitating flow characteristics in an axial flow waterjet pump. *Ocean Eng.* **2020**, *209*, 107450. [CrossRef]
38. Magrini, A.; Benini, E.; Yao, H.D.; Postma, J.; Sheaf, C. A review of installation effects of ultra-high bypass ratio engines. *Prog. Aerosp. Sci.* **2020**, *119*, 100680. [CrossRef]
39. Magrini, A.; Buosi, D.; Benini, E.; Sheaf, C. Ultra-high bypass nacelle geometry design space exploration. In Proceedings of the AIAA SciTech Forum 2021, Virtual, 11–15 & 19–21 January 2021. [CrossRef]
40. Magrini, A.; Buosi, D.; Benini, E. Maximisation of installed net resulting force through multi-level optimisation of an ultra-high bypass ratio engine nacelle. *Aerosp. Sci. Technol.* **2021**, *119*, 107169. [CrossRef]
41. Vuruskan, A.; Hosder, S. Impact of Turbulence Models and Shape Parameterization on Robust Aerodynamic Shape Optimization. *J. Aircr.* **2019**, *56*, 1099–1115. [CrossRef]
42. Ansys, Inc. *Ansys Fluent User's Guide*; Ansys Inc.: Canonsburg, PA, USA, 2022.
43. Menter, F.R. Two-equation eddy-viscosity turbulence models for engineering applications. *AIAA J.* **1994**, *32*, 1598–1605. [CrossRef]
44. Zwart, P.J.; Gerber, A.G.; Belamri, T. A two-phase flow model for predicting cavitation dynamics. In Proceedings of the Fifth International Conference on Multiphase Flow, Yokohama, Japan, 30 May–4 June 2004; Volume 152.
45. Ansys, Inc. *ANSYS Fluent Theory Guide*; Ansys Inc.: Canonsburg, PA, USA, 2022.
46. 24th I.T.T.C. The specialist committee on validation of waterjet test procedures: Final Report and Recommendations to the 24th ITTC. In Proceedings of the 24th International Towing Tank Conference, Edinburgh, UK, 4–10 September 2005.
47. Magrini, A.; Buosi, D.; Benini, E. Analysis of installation aerodynamics and comparison of optimised configuration of an ultra-high bypass ratio turbofan nacelle. *Aerosp. Sci. Technol.* **2022**, *128*, 107756. [CrossRef]

48. Celik, I.B.; Ghia, U.; Roache, P.J.; Freitas, C.J. Procedure for estimation and reporting of uncertainty due to discretization in CFD applications. *J. Fluids Eng. Trans. ASME* **2008**, *130*, 078001. [CrossRef]
49. Sobol', I. On the distribution of points in a cube and the approximate evaluation of integrals. *USSR Comput. Math. Math. Phys.* **1967**, *7*, 86–112. [CrossRef]

Journal of
Marine Science
and Engineering

MDPI

Article

Numerical Study of the Hydrodynamic Performance of a Two-Propeller Configuration

Xunming Wang, Yongyan Ma, Dakui Feng *, Chaobang Yao and Zichao Cai

Key Laboratory of Ship and Ocean Hydrodynamics of Hubei Province, School of Naval Architecture and Ocean Engineering, Huazhong University of Science and Technology, Wuhan 430074, China; wang_xm@hust.edu.cn (X.W.); mayongyan@hust.edu.cn (Y.M.); yaochaobang@hust.edu.cn (C.Y.); m202372543@hust.edu.cn (Z.C.)

* Correspondence: feng_dk@hust.edu.cn

Abstract: Improved ship design and market demands have driven the adoption of multi-propeller systems for propulsion in recent years. This study examines the hydrodynamic performance of two KP505 propellers arranged in various transverse and longitudinal spacings, utilizing an in-house CFD code. The numerical simulations employ the URANS method with the SST k-ω turbulence model and a structured overset grid approach. First, standardized mesh and time-step convergence studies are conducted following ITTC recommendations. The hydrodynamic results for the KP505 propeller are compared with experimental data to validate the reliability of the method. Subsequently, over 40 propeller arrangements with varying transverse and longitudinal spacing are simulated. Thrust, torque, and efficiency under different operating conditions are calculated, and key flow field data are analyzed. Finally, the interference characteristics between propellers at different positions are examined by comparing the results with those of a single KP505 propeller. The findings indicate that the high-speed wake generated by the upstream propeller significantly affects the hydrodynamic performance of the downstream propeller. This interaction diminishes as the transverse spacing between the propellers increases. To ensure the propulsion efficiency of the two-propeller configuration, the transverse spacing should not be less than one times the diameter of the propeller.

Keywords: flow field disturbance; multi-propeller interference; optimum propeller arrangement; hydrodynamic performance

1. Introduction

With advancements in ship design and increasing market demands, both the speed and tonnage of vessels have risen in recent years, placing greater demands on propulsion capabilities. However, the diameter of propellers is constrained by the available stern space and draft depth, making conventional single propellers increasingly inadequate to meet these propulsion needs. As a result, multi-propeller propulsion systems, including tandem and counter-rotating propellers, as well as twin-, triple-, and four-propeller configurations, have become more prevalent. Propellers in operation disrupt the surrounding flow field, which subsequently impacts the performance of other propellers within this disturbed environment. Therefore, propeller interference remains a critical issue in multi-propeller systems, as the interaction between the propellers may negatively impact their performance and ultimately reduce overall propulsion efficiency.

Tandem and counter-rotating propellers represent non-conventional marine propulsion systems that consist of two coaxial conventional propeller units. The key distinction between these configurations lies in their rotational dynamics: tandem systems maintain identical rotation direction for both propellers, while counter-rotating systems utilize opposite rotations. Both configurations exhibit minimal blade spacing in both transverse and longitudinal directions, resulting in relatively intense hydrodynamic interactions between the front and rear propellers.

Djahida et al. [1] studied the effects of axial displacement between tandem propellers. Their results indicated that with an appropriate configuration of the upstream and downstream propellers, it could be possible to achieve double or even greater thrust compared to a single propeller, while maintaining maximum efficiency. Güngör et al. [2] computed the acoustic performance of a counter-rotating propeller system (propeller types 3684 and 3685) using large eddy simulation combined with the FW-H method. The amplitudes of the pressure fluctuations showed good agreement with experimental data, excluding minor discrepancies in the third harmonic, demonstrating that the FW-H approach provided reliable results for the spectrum at both near- and far-field locations. Cui et al. [3] experimentally analyzed scour patterns of externally and internally rotating twin propellers using ADV measurements. Their findings revealed distinct scour concentration zones: maximum erosion for external rotation aligned with the individual propeller axes, while internal rotation induced peak scour along the coaxial midline between the propellers. Liu et al. [4] also found that by adjusting the spacing or diameter ratio, the downstream propeller could effectively absorb energy from the vortices shed by the upstream propeller, resulting in an approximately 45% improvement in the efficiency of the tandem propeller compared to a single propeller. Yao et al. [5,6] investigated the effects of axial distance and angular displacement on the hydrodynamic performance of tandem propellers based on RANS method. They observed that propeller efficiency could be enhanced at minimal axial distances combined with a specific angular offset. This phenomenon was attributed to favorable flow field interference between the front and rear propellers. They also found that the tandem propeller system has more advantages and potential for improving cavitation effects compared to conventional propellers. Chavan et al. [7] simulated the hydrodynamic performance of tandem propellers with different arrangements using a RANS CFD solver. They found that the resultant thrust and torque coefficients of the co-rotating propeller system increased compared to an equivalent single propeller, although the open-water efficiency decreased under moderate to high propeller load conditions. Pereira et al. [8] and Capone et al. [9] employed PIV to analyze near-wake vorticity dynamics in counter-rotating propellers. They discovered that the flow dynamics within a counter-rotating propeller system were characterized by the interaction of vorticity generated by both the front and rear propellers, which was dependent on blade loading and the relative phase angle. Their findings demonstrated that optimized operation of the aft propeller under specific loads effectively balanced the tangential velocity components. Jiang et al. [10] examined the hydrodynamic performance of a rim-driven counter-rotating thruster using the Ansys Fluent software. The results indicated that its hydrodynamic performances were better than those of a single-propeller rim-driven thruster, thereby confirming the advantages of this novel shaftless thruster.

For multi-propeller propulsion systems, particularly those with four propellers, the spacing between the propellers is more flexible and variable compared to the constrained spacing found in tandem and counter-rotating propeller arrangements. Both the transverse and longitudinal configurations can be adjusted between the two propellers on each side, coupled with the influence of the wake. As a result, the interaction characteristics among the propellers on multi-propeller ships can be quite complex. Optimizing the configuration of multi-propeller systems has become a crucial area of research.

Cozijn et al. [11] conducted thruster-interaction model tests in the Deepwater Towing Tank at MARIN. The PIV technique is applied to measure the wake flow of the azimuthing thrusters on a semi-submersible and a drill ship. The results indicated that the 'tilted' thrusters design can reduce thruster–thruster interaction effects. Bi et al. [12] presented propeller configuration effects on four-propeller vessel performance through systematic self-propulsion experiments. According to their 15 self-propulsion tests transverse positioning of inner propellers and longitudinal alignment of outer propellers emerged were the dominant factors governing the propulsion efficiency. Zong et al. [13] focused on the factors causing load changes in a four-propeller vessel. Their research found that pure interference from the other propeller contributes approximately 10% to total load variation, with inner propellers experiencing positive load increments and outer propellers exhibiting negative modulation. Sun et al. [14,15] conducted computational studies and experimental investigations on the propulsion performance of a four-screw ship. They examined the nominal wake and the distribution patterns of the wake field between the propeller disks. The results showed that the circumferential distribution patterns of the axial and tangential wake fields between the disks changed significantly depending on the positions of the inside and outside propellers. The mean axial velocity of the inside propeller was generally lower than that of the outside propeller. Zhou et al. [16] investigated thruster–thruster interactions in the vector-layout propulsion system of a remotely operated vehicle. They found that when the two thrusters were configured in tandem, the maximum thrust was achieved when the deviation angle between them was approximately 15°. Additionally, the results highlighted transverse spacing as the dominant factor influencing individual thruster performance compared to longitudinal variations. Yiew et al. [17] studied thruster and hull interactions on an azimuthing stern-drive tug through a series of model tests and numerical simulations. Their results demonstrated that propeller interactions predominantly influenced longitudinal thruster forces, whereas lateral forces stemmed primarily from thruster–unit drag, exhibiting minimal sensitivity to inter-thruster effects.

While single-propeller configurations have been extensively studied, covering aspects such as propeller loads during ship sailing and turning [18,19], propeller behavior under challenging off-design operational conditions [20,21], scale effects on hydrodynamic performance [22], wake evolution dynamics [23], and propeller cavitation [24], research on multi-propeller systems remains comparatively limited. Existing studies on multi-propeller configurations have either focused narrowly on specific propeller arrangements or prioritized propeller-hull interaction effects. Consequently, there are relatively few studies that specifically examine the transverse and longitudinal distribution of positions, as well as the characteristics of the interacting flow field in multi-propeller configurations, which significantly affect their hydrodynamic performance. In this paper, numerical simulations are performed using an in-house viscous CFD code to investigate the interactions within a two-propeller configuration. The simulations employ the URANS method with the SST k-ω turbulence model and a structured overset grid approach. Firstly, standardized verification and validation studies are conducted to demonstrate the reliability of the numerical method. Subsequently, two KP505 propellers are simulated in 45 different arrangements, with transverse and longitudinal spacing ranging from 0 to 1.6 and 0.5 to 4 times the propeller diameter, respectively. The effects of these arrangements on the hydrodynamic performance of both the front and rear propellers, as well as the characteristics of the flow field interactions, are analyzed. Finally, by comparing the propulsion efficiency of the two-propeller configuration with that of a single propeller, the optimal arrangement for two propellers in a multi-propeller system is proposed.

The paper is organized as follows: Section 2 provides a detailed description of the propeller model, numerical theories, computational domain, and grid distribution employed in the present study. In Section 3, the numerical method is verified through standard grid/time-step convergence studies, and it is validated against relevant experimental data. Section 4 outlines the simulation conditions and presents the simulation results for various transverse and longitudinal spacings, along with a discussion of the interaction characteristics between the two propellers. Finally, the conclusions are drawn in Section 5.

2. Materials and Methods

2.1. The Geometry of the KP505 Propeller

As shown in Figure 1, two identical KP505 propellers are selected as the simulation models, the open-water experimental data of KP505 can been found in the Tokyo CFD Workshop (2015) [25]. The details of the KP505 propeller are presented in Table 1.

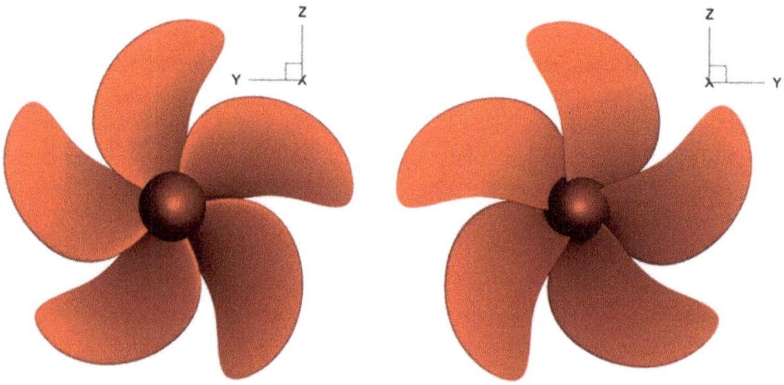

Figure 1. The geometry of KP505 propeller.

Table 1. The details of KP505 propeller.

Description	Values
Blade section	NACA66
Diameter D (m)	0.25
Pitch ratio P/D (0.7R)	0.95
Disk expanded area ratio Ae/Ao	0.8
Number of blades	5
Hub diameter ratio	0.18
Rotate speed (rps)	14
Rotation direction	Right-handed

2.2. Numerical Method

URANS and continuity equations are employed as the governing equations, which can be defined as:

$$\frac{\partial \rho}{\partial t} + \frac{\partial(\rho u_i)}{\partial x_i} = 0 \tag{1}$$

$$\frac{\partial u_i}{\partial t} + \frac{\partial u_i u_j}{\partial x_j} = -\frac{1}{\rho}\frac{\partial p}{\partial x_i} + \frac{1}{\rho}\frac{\partial}{\partial x_j}\left(\mu \frac{\partial u_i}{\partial x_j} - \rho \overline{u_i' u_j'}\right) \tag{2}$$

where u_i and u_i' (i = 1, 2, and 3) are the time averaged and fluctuation velocities component, respectively; ρ is the density; p is the time averaged pressure; μ is the dynamic viscosity and $\rho\overline{u_i' u_j'}$ is the Reynolds stress term. To compute the Reynolds stress term and close the governing equations, a two-equation SST $k-\omega$ turbulent model, proposed by Menter [26], is applied as the turbulence model. The model has advantages in predicting adverse pressure gradients and demonstrates good robustness to inlet conditions. The in-house

CFD code discretizes the governing equations using the finite difference method, and the PISO algorithm is employed to solve the velocity-pressure coupling equation [27]. Details of the numerical method can be found in the research [28,29].

2.3. Computational Domain and Boundary Conditions

As shown in Figure 2, the computational domain extends from $-6D$ to $12D$ along the X-axis, $-6D$ to $6D$ along the Y-axis, and $-6D$ to $6D$ along the Z-axis (where D is the diameter of the propeller). The inlet and outlet along the X-axis are set as velocity inlet and pressure outlet boundary conditions, respectively. The two side boundaries and the top are assigned as zero pressure gradient conditions, while the bottom of the domain is set as a zero pressure condition.

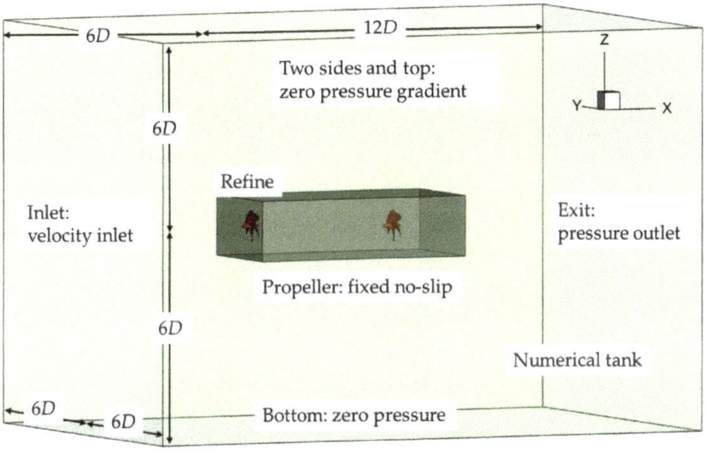

Figure 2. Computational domain and boundary conditions.

2.4. Overset Grid Method

The dynamic overset grid method is applied in the simulations. The main procedures of the in-house overset grid code include hole-digging, point-searching, and interpolation. The hole-mapping method [30] coupled with the cut-paste algorithm [31] is used to identify the hole nodes and remove unnecessary cells prior to computation. The ADT algorithm is employed to search for the donor nodes of the interpolation points. Computational information between the donor points and interpolation points is obtained using the trilinear interpolation method [32]. The dynamic overset grid technique has been employed in numerous propeller-related simulations [33,34].

The structured overset grid method is applied for both grid generation and propeller rotation simulations. As shown in Figure 3, the computational grids consist of background grids, refined grids, and model grids. To better capture the detailed flow field around the propeller, the refinement regions are carefully defined to suit different simulation conditions, details of the mesh distribution are illustrated in Figure 4.

The near-wall mesh height is set appropriately to meet the requirements of the chosen turbulence models, as shown in Figure 5, the nondimensional wall distance Y^+ distribution on the propeller surface is generally less than 0.5.

Both surface and volume grid overset techniques are applied, as shown in Figure 6.

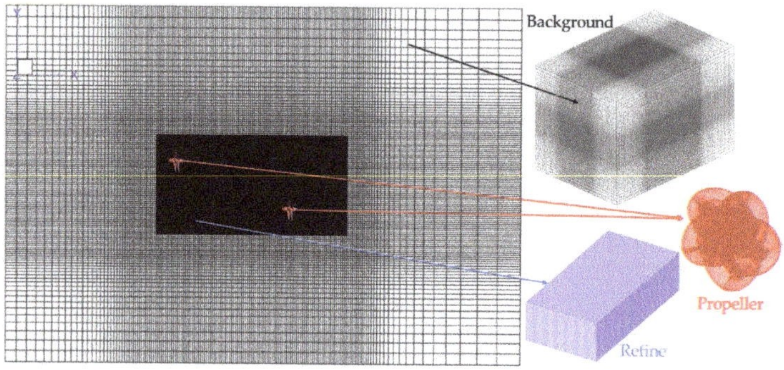

Figure 3. Different components of grids.

Figure 4. Grid distribution of different components.

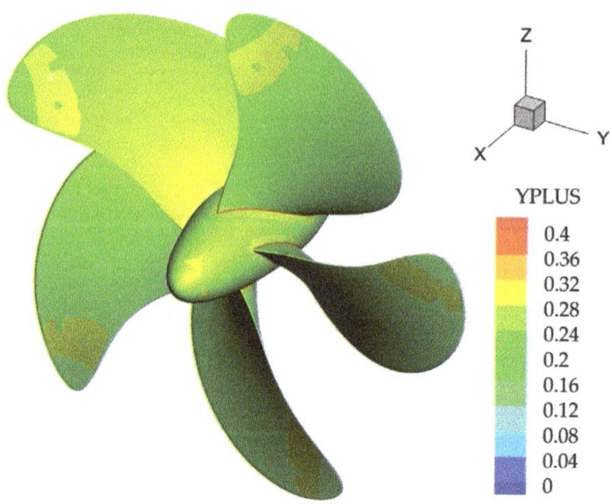

Figure 5. Y^+ on the propeller surface.

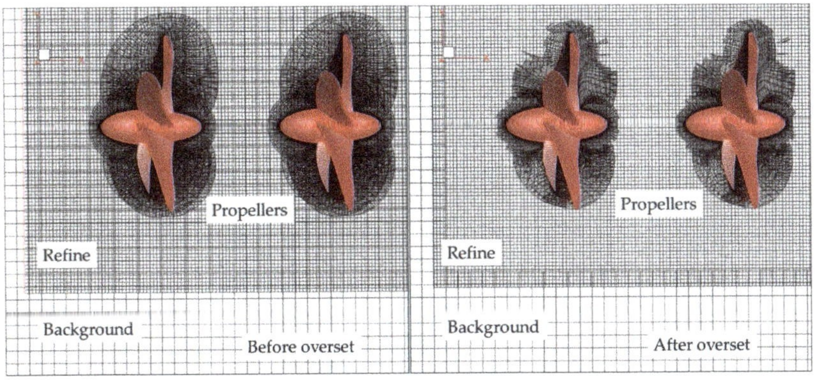

(a) (b)

Figure 6. Grids before and after overset: (**a**) Before overset; (**b**) After overset.

3. Verification and Validation

Before conducting the multi-propeller simulations, the present numerical method has been verified and validated. In this study, the thrust and torque coefficients for the propeller model are analyzed. The following coefficients are defined:

Advance coefficient:

$$J = \frac{U_0}{nD} \tag{3}$$

Thrust and torque coefficient:

$$K_T = \frac{T}{\rho n^2 D^4}, K_Q = \frac{Q}{\rho n^2 D^5} \tag{4}$$

Open-water efficient:

$$\eta = \frac{K_T}{K_Q}\frac{J}{2\pi} \tag{5}$$

where U_0 is the inflow velocity, n is the rotation speed, T is the thrust, and Q is the torque.

And the pressure coefficient C_P is defined as:

$$C_P = \frac{P_d}{0.5\rho U_0^2} \tag{6}$$

where P_d is the dynamic pressure.

The working condition of J = 0.8 for the KP505 propeller has been chosen for the numerical reliability analysis. According to the verification and validation (V&V) guidelines established by ASME [35,36], the V&V procedures mainly consist of two parts: verification for numerical uncertainty and validation based on experimental data.

3.1. Verification of the Numerical Uncertainty

Verification is defined as a procedure for assessing numerical uncertainty U_{SN}, which consists of the iteration uncertainty U_I, time step uncertainty U_T, and the grid space uncertainty U_G. The in-house CFD code has been validated through numerous simulations [37,38], and the iteration uncertainty is considered negligible. Therefore, two main aspects of numerical uncertainty are addressed here: the time step uncertainty U_T and the grid space uncertainty U_G. The numerical uncertainty U_{SN} is defined as:

$$U_{SN}{}^2 = U_G{}^2 + U_T{}^2 \tag{7}$$

According to the convergence study method proposed by Stern et al. [39] and Xing et al. [40], a minimum of three solutions in different grids and time-steps should be used in the convergence studies. Three sets of grids are shown in Figure 7 and Table 2, and the refinement ratio r_k between solutions is assumed as:

$$r_k = \frac{\Delta x_2}{\Delta x_1} = \frac{\Delta x_3}{\Delta x_2} \quad (8)$$

where Δx_i (i = 1, 2, and 3) are the coarse, medium, and fine grids or time steps, respectively.

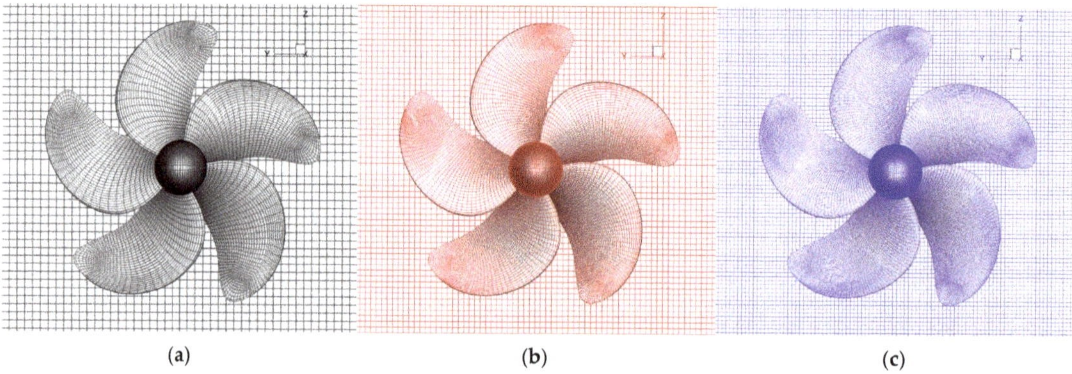

Figure 7. Three sets of grids: (**a**) Coarse grids; (**b**) Medium grids; (**c**) Fine grids.

Table 2. Dimensions of three grid cases.

Grid Nodes (M [1])	Propeller	Refine	Background	Total
Coarse	0.55	0.44	0.42	1.41
Medium	1.51	1.26	1.19	3.96
Fine	4.17	3.55	3.33	11.05

[1] M: million.

And S_i are the results corresponding to Δx_i. The changes between different solutions are $\varepsilon_{32} = S_3 - S_2$ and $\varepsilon_{21} = S_2 - S_1$. The convergence ratio R is defined as:

$$R = \frac{\varepsilon_{32}}{\varepsilon_{21}} \quad (9)$$

There are three convergence situations according to different R values:

(1) Monotonic convergence (MC): $0 < R < 1$
(2) Oscillatory convergence (OC): $R < 0$
(3) Divergence: $R > 1$

When $0 < R < 1$, monotonic convergence is achieved, the Richardson extrapolation method can be applied to estimate the numerical error δ_{RE} and order of accuracy P_{RE}:

$$\delta_{RE} = \frac{\varepsilon_{32}}{{r_k}^{P_{RE}} - 1} \quad (10)$$

$$P_{RE} = \frac{\ln(\varepsilon_{21}/\varepsilon_{32})}{\ln r_k} \quad (11)$$

The distance metric P_{DM} is given as:

$$P_{DM} = \frac{P_{RE}}{P_{th}} \quad (12)$$

where P_{th} = 2. Eventually, the factor of safety method is acquired as:

$$U_{FS} = \begin{cases} (2.45 - 0.85P_{DM})|\delta_{RE}|, 0 < P_{DM} \leq 1 \\ (16.4P_{DM} - 14.8)|\delta_{RE}|, P_{DM} > 1 \end{cases} \quad (13)$$

The coarse, medium, and fine grids consist of 1.41 million, 3.96 million, and 11.05 million grid nodes, respectively. Time steps are set to 0.004, 0.002 and 0.001 s, corresponding to 4, 2 and 1 degrees of propeller rotation per time step, respectively. The refinement ratio $r_k = \sqrt{2}$ for grid spacing and $r_k = 2$ for time steps are selected to estimate the numerical uncertainty. The KP505 simulations with different grids and time steps are based on the medium time step and grid nodes, respectively. The results are listed in Tables 3 and 4.

Table 3. Thrust and torque coefficients results of different grids.

	Grid Nodes (M [1])	K_T	Error	K_Q	Error
Coarse	1.41	0.131	−4.38%	0.0237	−4.05%
Medium	3.96	0.134	−2.19%	0.0242	−2.02%
Fine	11.05	0.135	−1.46%	0.0244	−1.21%
EFD	-	0.137	-	0.0247	-

[1] M: million.

Table 4. Thrust and torque coefficients results of different time steps.

	Δt (s)	K_T	Error	K_Q	Error
Coarse	0.004	0.127	−7.30%	0.0235	−4.86%
Medium	0.002	0.134	−2.19%	0.0242	−2.02%
Fine	0.001	0.136	−0.73%	0.0245	−0.81%
EFD	-	0.137	-	0.0247	-

As shown in Table 5, the uncertainties of the propeller thrust and torque for the grid spacing are 4.085%E and 3.714%E, while those for the time step are 0.982%E and 1.759%E (E denotes the experimental data). Both the propeller thrust and torque are more sensitive to grid spacing.

Table 5. Thrust and torque coefficients numerical uncertainties results.

		r_k	R	P_{DM}	U_{FS} (%E)	U_{SN} (%E)
K_T	Grid	$\sqrt{2}$	0.333	1.585	4.085	4.202
	Time step	2	0.286	0.904	0.982	
K_Q	Grid	$\sqrt{2}$	0.400	1.322	3.714	4.109
	Time step	2	0.429	0.611	1.759	

3.2. Validation Based on Experimental Data

The error W is defined as the difference between experimental data E and numerical results S:

$$W = E - S \quad (14)$$

And the validation uncertainty U_V is given by:

$$U_V{}^2 = U_E{}^2 + U_{SN}{}^2 \quad (15)$$

The comparison error W is compared with the validation uncertainty U_V, if $|W| < U_V$, then the validation is considered to have been achieved at the U_V level. Since the experimental uncertainty U_D of the thrust and torque of the KP505 propeller cannot be found in the reference, it is assumed to be 1.00%E. As shown in Table 6, the computation results are validated at the U_V level.

Table 6. Thrust and torque coefficients validation study results.

	U_{SN} (%E)	U_E (%E)	U_V (%E)	W (%E)
K_T	4.202	1.00	4.319	2.19
K_Q	4.109	1.00	4.229	2.02

4. Results and Discussion

4.1. Open-Water Characteristics of KP505 Propeller

The open-water performance of the KP505 propeller is computed and analyzed prior to investigating the flow interaction between two propellers. The simulation results and experimental data are presented in Table 7.

Table 7. Comparison of the K_T, $10K_Q$, and η between CFD and EFD data.

J	K_T			$10K_Q$			η		
	EFD	CFD	Error	EFD	CFD	Error	EFD	CFD	Error
0.1	0.482	0.470	2.49%	0.677	0.671	0.89%	0.113	0.112	1.30%
0.2	0.435	0.436	−0.23%	0.622	0.626	−0.64%	0.223	0.222	0.53%
0.3	0.387	0.393	−1.55%	0.557	0.573	−2.87%	0.332	0.328	1.31%
0.4	0.336	0.345	−2.68%	0.497	0.514	−3.42%	0.431	0.428	0.81%
0.5	0.285	0.294	−3.16%	0.437	0.452	−3.43%	0.519	0.518	0.22%
0.6	0.235	0.242	−2.98%	0.376	0.387	−2.93%	0.597	0.597	−0.07%
0.7	0.185	0.188	−1.62%	0.311	0.316	−1.61%	0.665	0.663	0.28%
0.8	0.137	0.134	2.19%	0.247	0.242	2.02%	0.705	0.705	−0.05%
0.9	0.083	0.078	6.02%	0.181	0.166	8.29%	0.654	0.673	−2.97%

Except for J = 0.9, which exhibits relatively large discrepancies between CFD and EFD data, the errors for the other advance coefficients remain within 4%. The open-water characteristics curve is plotted in Figure 8. Overall, the computational results align well with the EFD data.

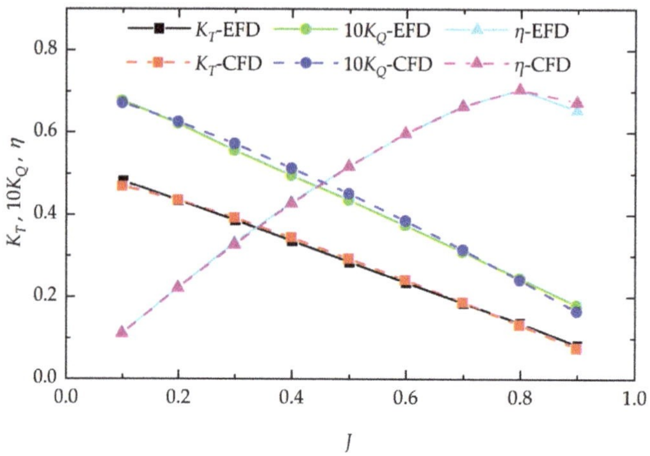

Figure 8. Open-water characteristics of KP505 propeller.

The advance coefficient J = 0.8 (corresponding to U_0 = 2.8 m/s, n = 14 rps, and D = 0.25 m), utilized in Verification and Validation, has been chosen for the subsequent analysis of propeller interference in the two-propeller configuration.

4.2. Simulation Conditions of the Two-Propeller Configuration

To investigate the effect of various arrangements on the hydrodynamic performance of two KP505 propellers, a total of 45 simulation conditions have been established.

The transverse spacing (*TS*) ranges from 0 to 1.6*D*, while the longitudinal spacing (*LS*) varies from 0.5*D* to 4*D* (where *D* represents the propeller diameter). The spacing between the front and rear propellers in both the transverse and longitudinal directions is defined in Figure 9. The specific layout is detailed in Table 8.

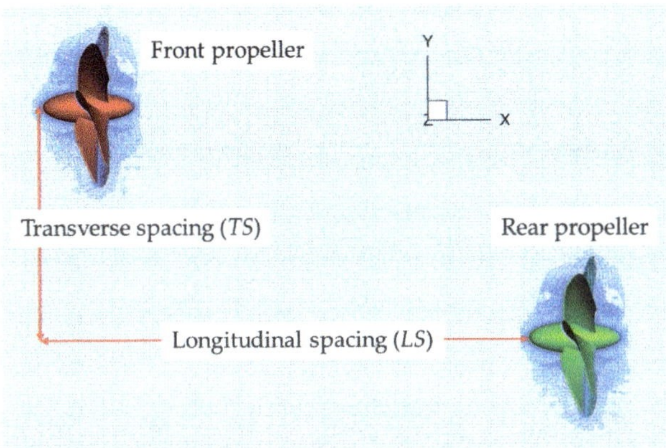

Figure 9. Definition of the transverse spacing (*TS*) and longitudinal spacing (*LS*).

Table 8. Propellers arrangement of different conditions.

Conditions		*LS* (*D*)				
		0.5	1	2	3	4
TS (*D*)	0	C1	C2	C3	C4	C5
	0.2	C6	C7	C8	C9	C10
	0.4	C11	C12	C13	C14	C15
	0.6	C16	C17	C18	C19	C20
	0.8	C21	C22	C23	C24	C25
	1.0	C26	C27	C28	C29	C30
	1.2	C31	C32	C33	C34	C35
	1.4	C36	C37	C38	C39	C40
	1.6	C41	C42	C43	C44	C45

4.3. The Effect of Different Layout on the Front Propeller

The thrust and torque coefficients of the front propeller under different simulation conditions are presented in Tables 9 and 10. It can be observed that the variations in the thrust and torque coefficients of the front propeller under different operating conditions are relatively minor. Among the 45 arrangements analyzed, the maximum differences in the thrust and torque coefficients of the front propeller are approximately 6% and 3% (C1), respectively, when compared to those of the single KP505 propeller. Except for cases at *LS* = 0.5*D* and 1*D*, where variations in K_T and K_Q under certain transverse spacings are more pronounced, the amplitudes of the thrust and torque coefficients remain largely consistent under other operating conditions.

Table 9. The K_T of the front propeller in different conditions.

K_T		LS (D)				
		0.5	**1**	**2**	**3**	**4**
TS (D)	0	0.142	0.137	0.134	0.134	0.134
	0.2	0.140	0.135	0.134	0.134	0.134
	0.4	0.133	0.133	0.133	0.133	0.134
	0.6	0.130	0.132	0.133	0.133	0.134
	0.8	0.129	0.131	0.133	0.133	0.134
	1.0	0.130	0.132	0.133	0.133	0.134
	1.2	0.131	0.132	0.133	0.134	0.133
	1.4	0.132	0.132	0.133	0.133	0.133
	1.6	0.133	0.133	0.133	0.133	0.133

Table 10. The $10K_Q$ of the front propeller in different conditions.

$10K_Q$		LS (D)				
		0.5	**1**	**2**	**3**	**4**
TS (D)	0	0.250	0.245	0.243	0.243	0.243
	0.2	0.248	0.244	0.242	0.242	0.243
	0.4	0.242	0.241	0.242	0.242	0.243
	0.6	0.238	0.240	0.242	0.242	0.243
	0.8	0.237	0.240	0.241	0.242	0.243
	1.0	0.238	0.240	0.241	0.242	0.242
	1.2	0.240	0.240	0.241	0.243	0.242
	1.4	0.241	0.241	0.241	0.242	0.242
	1.6	0.241	0.241	0.242	0.242	0.242

When $LS = 0.5D$, the blade surfaces of the front and rear propellers are in close proximity along the longitudinal direction. The flow field around the front propeller is disrupted by the rotational turbulence generated by the rear propeller.

The influence of the rear propeller can initially be positive but it may turn negative as the transverse spacing increases. As depicted in Figures 10 and 11, both the thrust and torque coefficients are greater at $TS = 0$ and $TS = 0.2D$, but lower at $TS = 0.6D$ and $TS = 0.8D$, compared to the values for a single KP505 propeller.

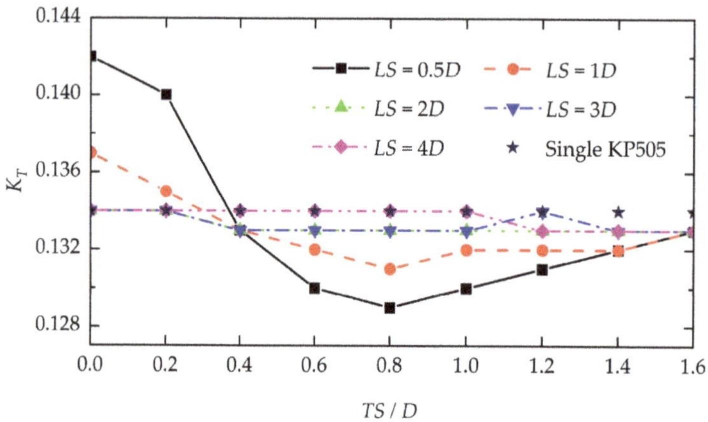

Figure 10. The K_T of the front propeller in different arrangements and single KP505 propeller.

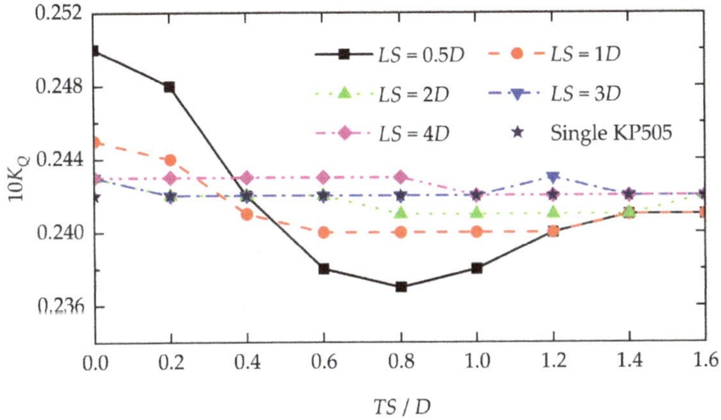

Figure 11. The $10K_Q$ of the front propeller in different arrangements and single KP505 propeller.

The axial velocity distributions at the front propeller plane under different operation-al conditions are presented in Figure 12. When influenced by the rear propeller, the axial velocity distribution on the front propeller plane exhibits distinct variations: at *TS* = 0, a relatively large reduction is observed compared to the single KP505 propeller, while at *TS* = 0.8*D*, the distribution demonstrates slightly enhanced axial velocity characteristics. The velocity improvement in Figure 12c is more evident in the right-side region of the propeller plane, where the flow field undergoes direct interaction with the rear propeller, as illustrated in Figure 13.

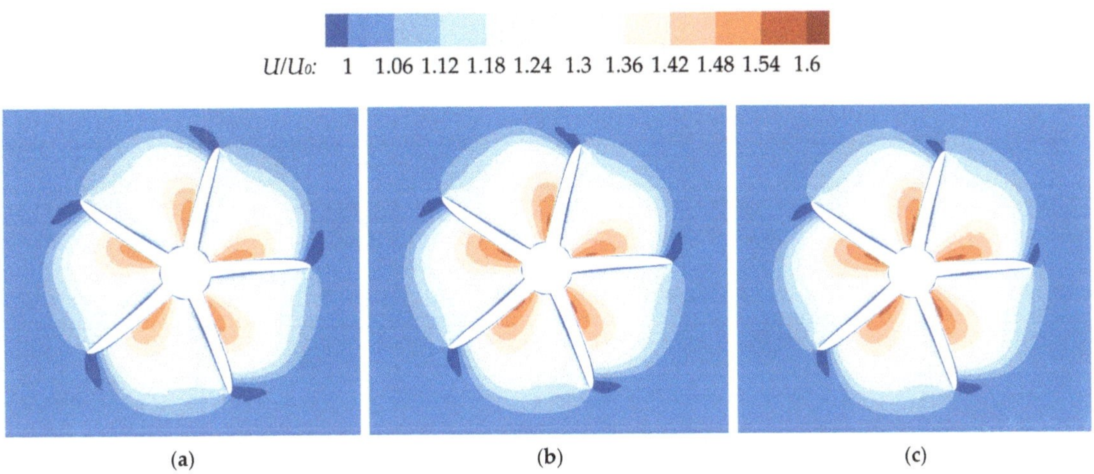

Figure 12. Comparison of axial velocity distributions between the front propeller plane under different conditions and a single propeller plane (**a**) C1 (*TS* = 0, *LS* = 0.5*D*); (**b**) Single propeller; (**c**) C21 (*TS* = 0.8*D*, *LS* = 0.5*D*).

Under a constant propeller rotation speed *n* and diameter *D*, the elevated axial velocity corresponds to an increased advance coefficient *J*, thereby reducing the hydrodynamic performance of the propeller. Conversely, a reduced axial velocity induces higher propeller loads. Therefore, both the thrust and torque coefficients increase at *TS* = 0 but decrease at *TS* = 0.8*D*. The corresponding coefficients at *TS* = 0 and *TS* = 0.8*D* achieve their maximum and minimum values, respectively. Specifically, the thrust coefficient of the front propeller surpasses the K_T of the single KP505 propeller by 5.97% at $TS = 0$, while diminishing by 3.73% at $TS = 0.8D$.

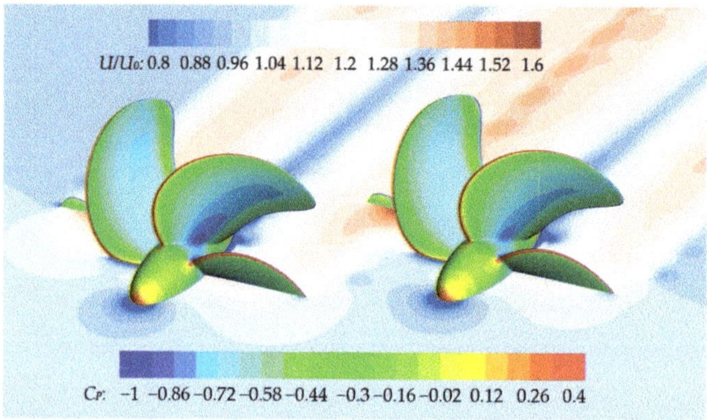

Figure 13. The axial velocity distribution on the XOY plane of the flow field at C21 ($TS = 0.8D$, $LS = 0.5D$).

Furthermore, alterations in axial velocity distribution correspondingly modify the surface pressure characteristics of the front propeller. At $TS = 0$, these pressure distributions exceed those observed in the single KP505 propeller configuration, whereas at $TS = 0.8D$, they demonstrate marked reduction. This comparative behavior is clearly illustrated in Figure 14. The effect of the rear propeller eventually becomes negligible when TS exceeds $1D$, with thrust and torque coefficient deviations remaining within 3% relative to the single KP505 propeller.

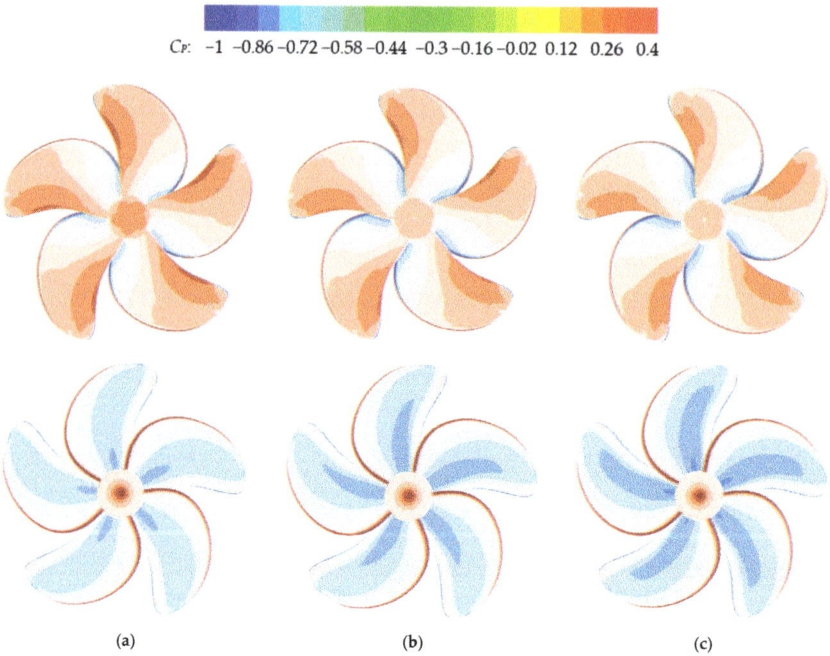

Figure 14. Pressure distributions on the front propeller surfaces at C1 and C21, compared to those of a single KP505 propeller: (**a**) C1 ($TS = 0$, $LS = 0.5D$); (**b**) Single propeller; (**c**) C21 ($TS = 0.8D$, $LS = 0.5D$).

When $LS = 1D$, the variation trends of the thrust and torque coefficients resemble those observed at $LS = 0.5D$. However, as the longitudinal distance between the propellers increases, the interaction effect from the rear propeller weakens. The maximum differences in thrust and torque coefficients compared to those of a single KP505 propeller are 2.24% and 1.24%, respectively. Consequently, the peak values of the thrust and torque coefficients are lower than those recorded at $LS = 0.5D$.

When $LS > 1D$, the interference from the rear propeller decreases significantly, and the thrust and torque coefficients of the front propeller are almost the same as those of a single propeller.

4.4. The Effect of Different Layout on the Rear Propeller

The thrust and torque coefficients of the rear propeller under different simulation conditions are presented in Tables 11 and 12, with the corresponding values plotted in Figures 15 and 16.

Table 11. The K_T of the rear propeller in different conditions.

K_T		LS (D)				
		0.5	**1**	**2**	**3**	**4**
TS (D)	0	−0.035	−0.037	−0.037	−0.036	−0.035
	0.2	0.018	0.013	0.010	0.009	0.010
	0.4	0.090	0.082	0.079	0.078	0.078
	0.6	0.116	0.111	0.108	0.107	0.107
	0.8	0.133	0.131	0.129	0.128	0.127
	1.0	0.137	0.137	0.135	0.134	0.134
	1.2	0.136	0.136	0.135	0.135	0.134
	1.4	0.135	0.135	0.135	0.134	0.134
	1.6	0.135	0.135	0.135	0.134	0.134

Table 12. The $10K_Q$ of the rear propeller in different conditions.

$10K_Q$		LS (D)				
		0.5	**1**	**2**	**3**	**4**
TS (D)	0	0.035	0.032	0.031	0.032	0.034
	0.2	0.101	0.099	0.096	0.095	0.098
	0.4	0.228	0.225	0.222	0.221	0.221
	0.6	0.265	0.263	0.261	0.260	0.260
	0.8	0.252	0.253	0.252	0.251	0.251
	1.0	0.245	0.245	0.244	0.244	0.244
	1.2	0.244	0.244	0.244	0.243	0.243
	1.4	0.244	0.244	0.243	0.243	0.242
	1.6	0.244	0.244	0.243	0.243	0.242

According to the data presented in Figures 15 and 16, variations in rear propeller thrust and torque coefficients with respect to different longitudinal spacings (ranging from 0.5D to 4D) are negligible when the transverse spacing is held constant. This insensitivity can be attributed to the minimal attenuation of flow velocity within the wake field of the front propeller as it propagates downstream, which can be observed from Figures 17–19.

Figure 17 indicates that the rear propeller experiences only slight perturbations in axial flow velocity at various longitudinal positions within 4D. Figures 18 and 19 demonstrate the axial and tangential velocity distributions on the rear propeller plane for $LS = 1D$ and 4D, respectively. It is evident that the axial velocity exhibits nearly identical distributions, while the tangential velocity shows only a slight decay between $LS = 1D$ and $LS = 4D$.

Therefore, the thrust and torque coefficients of the rear propeller remain largely unaffected by longitudinal spacing when *LS* is less than 4*D*.

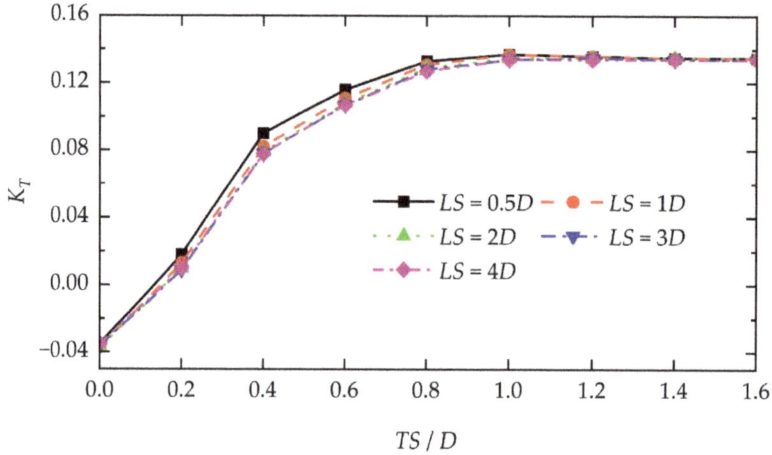

Figure 15. The K_T of the rear propeller in different arrangements.

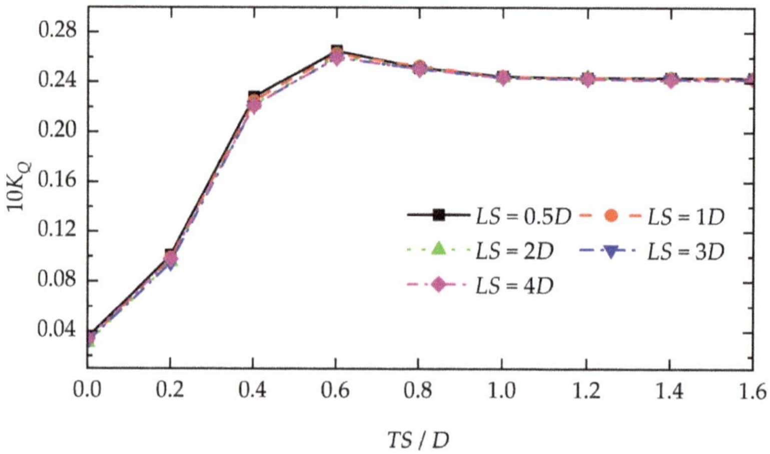

Figure 16. The $10K_Q$ of the rear propeller in different arrangements.

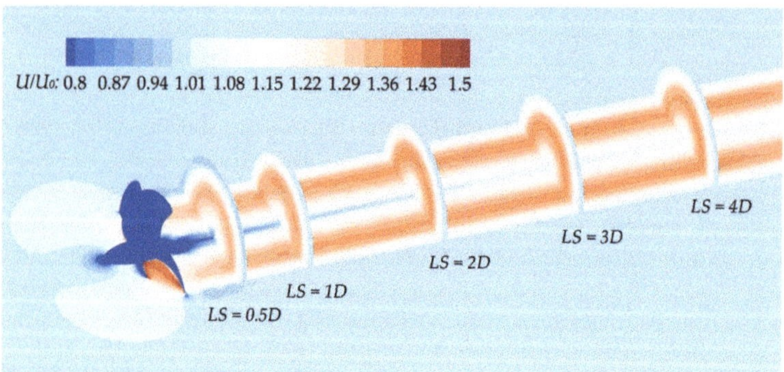

Figure 17. The axial velocity of the KP505 propeller wake at different longitudinal positions.

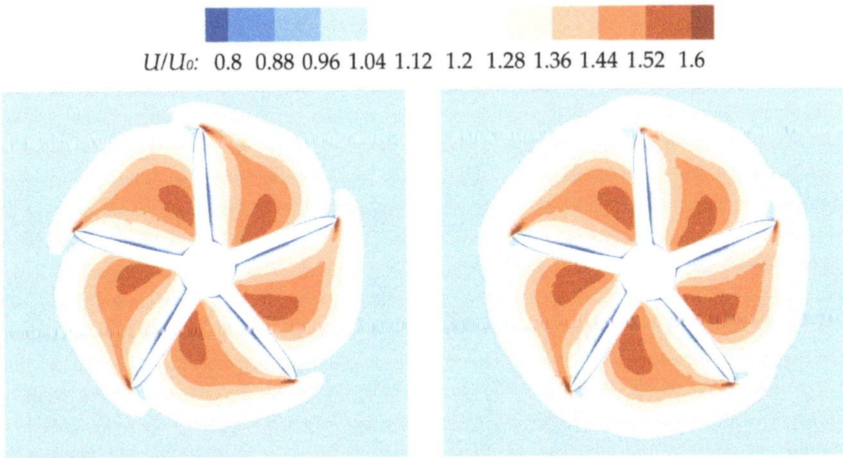

(a) (b)

Figure 18. The axial velocity distributions on the rear propeller plane at different longitudinal spacings: (**a**) C2 ($TS = 0$, $LS = 1D$; (**b**) C5 ($TS = 0$, $LS = 4D$).

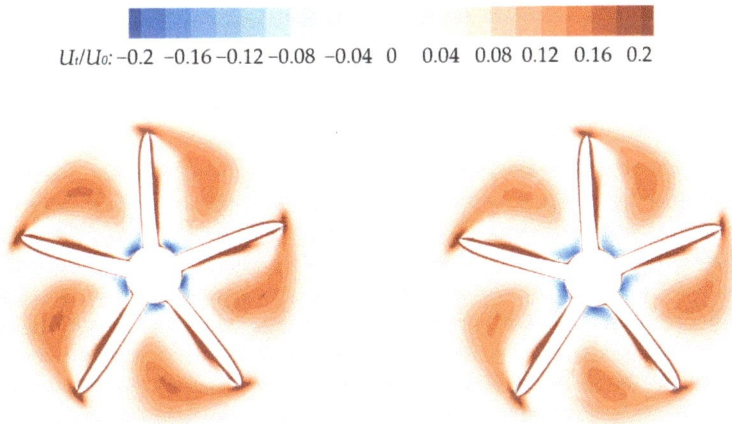

(a) (b)

Figure 19. The tangential velocity distributions on the rear propeller plane at different longitudinal spacings: (**a**) C2 ($TS = 0$, $LS = 1D$; (**b**) C5 ($TS = 0$, $LS = 4D$).

As for the different transverse spacings, when the transverse spacing between the two propellers is small ($TS < 0.6D$), the thrust and torque coefficients of the rear propeller are significantly reduced due to the interference from the front propeller. As shown in Figures 18–21, both the axial and tangential velocity distributions on the rear propeller plane at $TS = 0$ (Figures 18a and 19a) and $0.4D$ (Figures 20a and 21a) are considerably greater than those of a single KP505 propeller in Figures 20c and 21c. An increased onset flow velocity deteriorates the hydrodynamic performance of the rear propeller, as discussed in Section 4.3. Furthermore, according to blade element theory [37,41], the fundamental parameters are defined as follows:

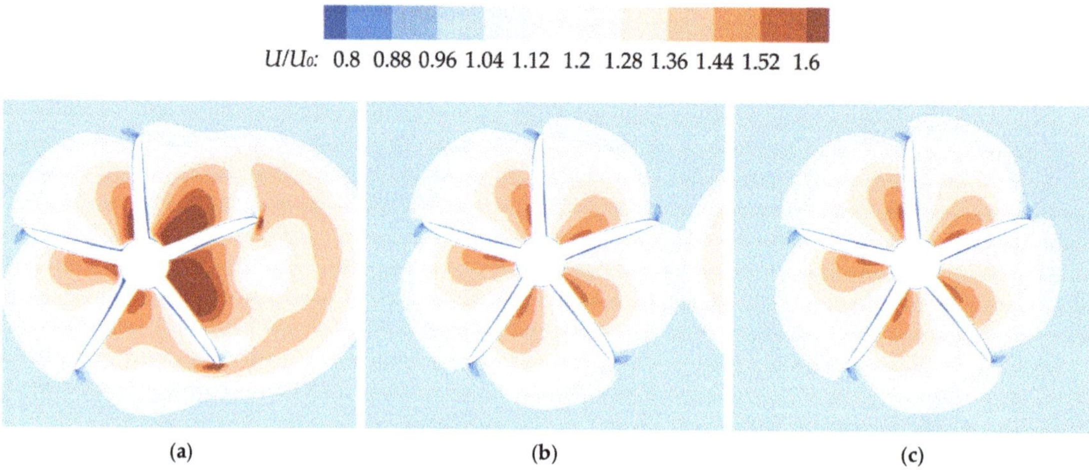

Figure 20. Comparison of axial velocity distributions between the rear propeller plane under different conditions and a single propeller plane (**a**) C12 (TS = 0.4D, LS = 1D); (**b**) C27 (TS = 1D, LS = 1D); (**c**) Single propeller.

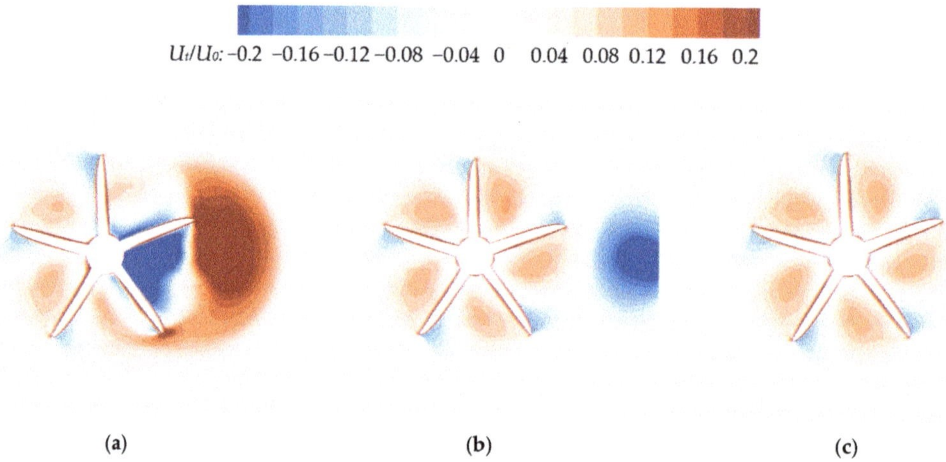

Figure 21. Comparison of tangential velocity distributions between the rear propeller plane under different conditions and a single propeller plane (**a**) C12 (TS = 0.4D, LS = 1D); (**b**) C27 (TS = 1D, LS = 1D); (**c**) Single propeller.

Hydrodynamic pitch angle β_i:

$$\beta_i = \arctan(\frac{U}{2\pi rn - U_t}) \quad (16)$$

Pitch angle θ:

$$\theta = \frac{P}{2\pi r} \quad (17)$$

The angle of attack α_K:

$$\alpha_K = \theta - \beta_i \quad (18)$$

where U (axial) and U_t (tangential) denote instantaneous velocities incorporating induced components derived from CFD results, with P representing the propeller pitch.

When the transverse spacing between propellers is minimized, both the axial velocity U and the tangential velocity U_t of the rear propeller are elevated compared to the single KP505 propeller. This increase induces an increased hydrodynamic pitch angle β_i, ultimately reducing the angle of attack α_K. Within the practical operational range (characterized by small angle of attack α_K values), the lift coefficient C_L exhibits approximate linear proportionality to the angle of attack α_K, while the drag coefficient $C_D \propto {C_L}^2$ [37,42]. The concurrent decrease of C_L and C_D, with diminishing α_K, directly correlates with observed reductions in thrust and torque coefficients under minimal transverse spacing conditions. Particularly at $TS = 0$, the rear propeller is entirely immersed within the high-velocity wake generated by the front propeller, as visualized in Figure 22a. This arrangement induces severely elevated inflow velocities, resulting in negative thrust coefficients for the rear propeller as demonstrated in Table 11 and Figure 15.

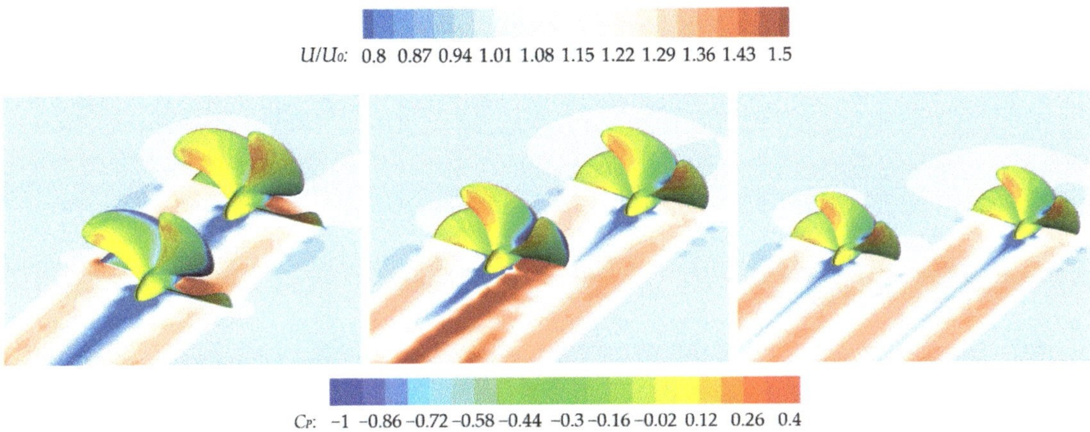

(a) (b) (c)

Figure 22. Comparison of the axial velocity distribution on the XOY plane and the pressure distribution on the propeller surface under different conditions (**a**) C2 ($TS = 0$, $LS = 1D$); (**b**) C12 ($TS = 0.4D$, $LS = 1D$); (**c**) C27 ($TS = 1D$, $LS = 1D$).

Figure 22b presents the flow field and pressure distribution on the rear propeller at condition C12 ($TS = 0.4D$). Due to partial interaction with the high-speed inflow from the front propeller, the axial velocity within the right-hand quadrant of the XOY section obviously exceeds values observed on the left side. Simultaneously, the pressure magnitudes on the right side demonstrate marked reduction compared to the left region. These asymmetric patterns align with the detailed velocity and pressure characteristics documented in Figures 20a and 23b.

Figure 23 presents the pressure distributions on the surface of the rear propeller under different conditions. Due to the high-speed axial inflow velocity, the pressure distributions on the rear propeller are higher on the suction side while lower on the pressure side (Figure 23a,b) compared to those of the single KP505 propeller (Figure 14b). The reduction in the pressure difference between the pressure and suction sides also accounts for the reduction in the hydrodynamic loads of the rear propeller.

The impact of the front propeller gradually reduced as the transverse distance between the two propellers increases. The interference from the front propeller becomes negligible while the transverse spacing exceeds $0.8D$. As illustrated in Figure 22c, the wake of the front propeller exhibits minimal interaction with the rear propeller at $TS = 1D$, both the axial velocity and pressure distributions of the two propellers are nearly identical. Furthermore, at $TS = 1D$, the velocity and pressure distributions of the rear propeller closely resemble those of the single KP505 propeller, as depicted in Figures 20b, 21b and 23c. Simultaneously, the thrust and torque coefficients of the rear propeller are almost identical to those of the single KP505 propeller.

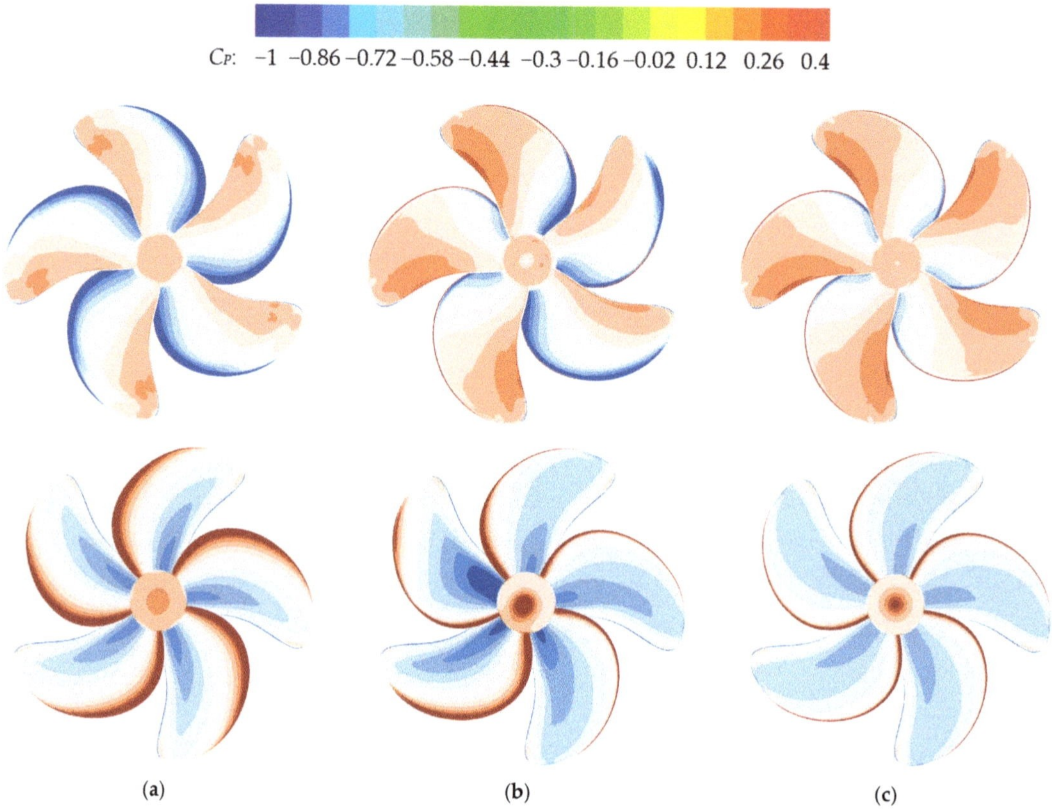

Figure 23. Pressure distribution on the rear propeller surface under different conditions: (**a**) C2 ($TS = 0$, $LS = 1D$); (**b**) C12 ($TS = 0.4D$, $LS = 1D$); (**c**) C27 ($TS = 1D$, $LS = 1D$).

4.5. The Effect of Different Layout on the Total Performance of the Two-Propeller Configuration

The total thrust and torque coefficients for the two-propeller configuration under various arrangements are shown in Figures 24 and 25. For the two-propeller configuration investigated in this study, the interaction effect is primarily manifested in the impact of the front propeller wake on the hydrodynamic performance of the rear propeller, while the rear propeller has a relatively small influence on the front propeller. Consequently, the variation trends of the total thrust and torque coefficients are almost identical to those of the rear propeller.

The comparison of propulsion efficiency between the two-propeller configuration in various arrangements and the single KP505 propeller is presented in Figure 26. The efficiency is also more sensitive to the transverse spacing, while the effect of the longitudinal spacing remains relatively constant. Propulsion efficiency increases with greater transverse spacing between the propellers.

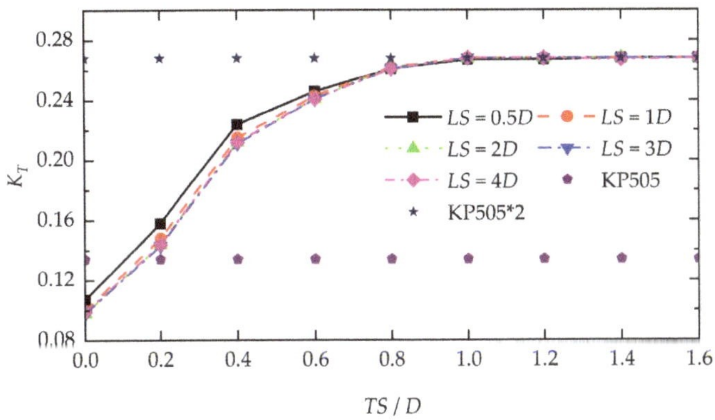

Figure 24. The total K_T of the two-propeller configuration in different arrangements.

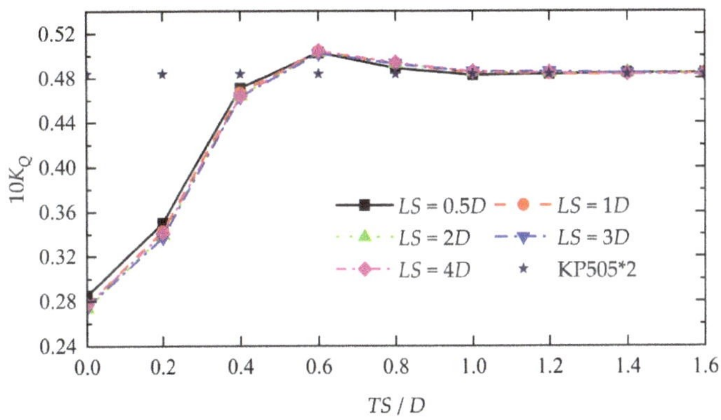

Figure 25. The total $10K_Q$ of the two-propeller configuration in different arrangements.

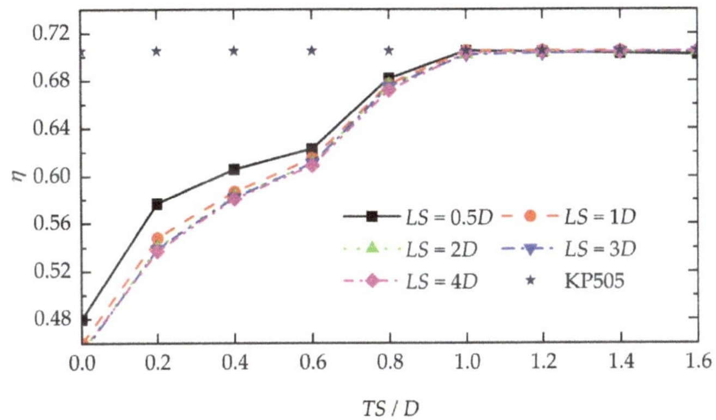

Figure 26. The total efficiency of the two-propeller configuration in different arrangements.

For transverse spacing values exceeding $TS = 0.2D$, the two-propeller configuration produces higher combined thrust than the single propeller configuration (Figure 24). However, to achieve double the thrust output of the single KP505 propeller, the transverse spacing of the two-propeller configuration should be at least $TS \geq 1D$. Notably, this system attains peak efficiency comparable to the single KP505 propeller when $TS \geq 1D$. Therefore, to ensure optimal thrust and efficiency of the two-propeller configuration, the transverse spacing between the two propellers should be maintained at no less than 1D.

5. Conclusions

In this paper, the interaction characteristics of a two-KP505 propeller configuration in various arrangements are studied using a URANS code. The hydrodynamic performance and flow field characteristics of the two propellers, with transverse spacing ranging from 0 to 1.6D and longitudinal spacing ranging from 0.5D to 4D, are calculated and analyzed.

The results indicate that, for the two-propeller configuration investigated in this study, the interaction between propellers is predominantly characterized by the high-speed wake generated by the upstream propeller, which significantly affects the hydrodynamic performance of the downstream propeller. In contrast, the effect of the downstream propeller on the upstream propeller is relatively minor.

For the front propeller, the performance of the front propeller is primarily influenced by the rear propeller when they are in close proximity. According to the results, limited interaction occurs when both the transverse and longitudinal spacing are less than one times the diameter of the propeller. The differences in the thrust and torque coefficients of the front propeller are all within 6% when compared to those of the single KP505 propeller.

For the rear propeller, the wake of the front propeller exhibits minimal velocity decay across longitudinal spacings ranging from 0.5D to 4D. Consequently, variations in longitudinal spacing within this range have insignificant effects on the hydrodynamic performance of the rear propeller. The rear propeller is more sensitive to changes in transverse spacing. When the transverse spacing between the two propellers is small, the rear propeller experiences substantial influence from the high-velocity inflow induced by the front propeller. This flow interaction induces three concurrent effects: (1) accelerated onset flow velocity across the propeller disk, (2) diminished blade angle of attack (α_K), and (3) reduced pressure difference on the propeller surfaces, collectively resulting in reductions in thrust output and torque generation. The influence of the front propeller progressively diminishes with increasing transverse spacing, becoming negligible when the spacing exceeds 0.8 times the diameter of the propeller.

The variation trend of the total thrust and torque coefficients for the two-propeller configuration closely resembles that of the rear propeller. Propulsion efficiency improves as the transverse spacing between the propellers increases. The two-propeller configuration can generate greater combined thrust than the single KP505 propeller while $TS > 0.2D$. However, to ensure optimal overall propulsion efficiency, the longitudinal spacing can be increased while maintaining a transverse spacing of not less than one times the propeller diameter.

For future studies, multiple advance coefficients will be considered, as the current research is based on a single $J = 0.8$. Thus, a more comprehensive investigation of the propeller interference characteristics under different arrangement forms for varying advance coefficients can be conducted.

Author Contributions: Data curation, X.W.; Formal analysis, D.F.; Funding acquisition, D.F. and C.Y.; Investigation, Z.C.; Methodology, D.F.; Project administration, Y.M.; Resources, D.F.; Software, X.W., D.F. and C.Y.; Supervision, D.F. and C.Y.; Validation, Y.M. and Z.C.; Visualization, X.W.; Writing—original draft, X.W.; Writing—review and editing, D.F. and C.Y. All authors have read and agreed to the published version of the manuscript.

Funding: This research was funded by the YEQISUN Joint Funds of the National Science Foundation of China (Grant No. U2141228), the National Natural Science Foundation of China (Grant No. 52071148), and Special Study on Civil aircraft of Ministry of Industry and Information Technology (MJZ5-3N21-1). The APC was funded by U2141228. The essential supports are greatly acknowledged.

Data Availability Statement: The original contributions presented in this study are included in the article. Further inquiries can be directed to the corresponding author.

Conflicts of Interest: The authors declare no conflicts of interest.

References

1. Djahida, B.; Omar, I. Impact of some geometrical aspects on the tandem co-rotating propeller hydrodynamic characteristics. *Brodogradnja* **2017**, *68*, 107–123. [CrossRef]
2. Güngör, E.; Özdemir, I.B. Prediction of Noise and Acoustical Spectrum of Counter-Rotating Propellers. *J. Ship Res.* **2018**, *62*, 166–182. [CrossRef]
3. Cui, Y.; Lam, W.H.; Robinson, D.; Hamill, G. Temporal and spatial scour caused by external and internal counter-rotating twin-propellers using Acoustic Doppler Velocimetry. *Appl. Ocean Res.* **2020**, *97*, 102093. [CrossRef]
4. Liu, Y.; Gong, Q. Numerical investigation on the flow characteristics and hydrodynamic performance of tandem propeller. *Appl. Ocean Res.* **2020**, *101*, 102292. [CrossRef]
5. Yao, H.; Liu, Y.; Zhang, H.; Zhang, Q. Numerical investigation on hydrodynamic performance of new canard-configuration tandem propellers. *Appl. Ocean Res.* **2020**, *104*, 102374. [CrossRef]
6. Yao, H.; Liu, Y.; Zhang, H.; Zhang, Q. Comparative study on hydrodynamic performance and induced pressure of new canard tandem propellers and conventional propellers. *Ocean Eng.* **2021**, *221*, 108566. [CrossRef]
7. Chavan, S.A.; Bhattacharyya, A.; Sha, O.P. Open water performance of B-Series marine propellers in tandem configurations. *Ocean Eng.* **2021**, *242*, 110158. [CrossRef]
8. Pereira, F.A.; Capone, A.; Felice, F.D. Flow field and vortex interactions in the near wake of two counter-rotating propellers. *Appl. Ocean Res.* **2021**, *117*, 102918. [CrossRef]
9. Capone, A.; Felice, F.D.; Pereira, F.A. On the flow field induced by two counter-rotating propellers at varying load conditions. *Ocean Eng.* **2021**, *221*, 108322. [CrossRef]
10. Jiang, H.; Wu, O.; Sheng, C.; Lan, J.; Bucknall, R. Numerical investigation on hydrodynamic performance of a novel shaftless rim-driven counter-rotating thruster considering gap fluid. *Appl. Ocean Res.* **2022**, *118*, 102967. [CrossRef]
11. Cozijn, J.L.; Hallmann, R. Thruster-Interaction Effects on a DP Semi-Submersible and a Drill Ship: Measurement and Analysis of the Thruster Wake Flow. In Proceedings of the ASME 2013 32nd International Conference on Ocean, Offshore and Arctic Engineering, Nantes, France, 9–14 June 2013.
12. Bi, J.Y.; Zong, Z. Model Test of Influence of Propeller Arrangements Aboard a Four Propeller Boat. *ASME. J. Offshore Mech. Arct. Eng.* **2013**, *135*, 021101. [CrossRef]
13. Zong, Z.; Hong, Z.; Zhang, H.; Sahoo, P.K. An investigation of load change of propeller in a four-propulsion vessel using CFD. *J. Mar. Sci. Technol.* **2018**, *23*, 122–131. [CrossRef]
14. Sun, S.; Chang, X.; Guo, C.; Zhang, H.; Wang, C. Numerical investigation of scale effect of nominal wake of four-screw ship. *Ocean Eng.* **2019**, *183*, 208–223. [CrossRef]
15. Sun, S.; Chang, X.; Wang, W.; Guo, C.; Zhang, H. Numerical simulation of wake field and propulsion performance in a four-screw ship. *Ocean Eng.* **2019**, *186*, 106092. [CrossRef]
16. Zhou, B.; Zhao, M. Numerical simulation of thruster-thruster interaction for ROV with vector layout propulsion system. *Ocean Eng.* **2020**, *210*, 107542. [CrossRef]
17. Yiew, L.J.; Taskar, B.; Jin, Y.; Zheng, Y.; Liu, D.; Magee, A.R. Experimental and numerical investigations on coupled thruster and hull interactions for a benchmark ASD tug. *Appl. Ocean Res.* **2022**, *120*, 103072. [CrossRef]
18. Ortolani, F.; Dubbioso, G. Experimental investigation of single blade and propeller loads by free running model test. Straight ahead sailing. *Appl. Ocean Res.* **2019**, *87*, 111–129. [CrossRef]
19. Ortolani, F.; Dubbioso, G. Experimental investigation of blade and propeller loads: Steady turning motion. *Appl. Ocean Res.* **2019**, *91*, 101874. [CrossRef]
20. Yurtseven, A.; Aktay, K. The numerical investigation of spindle torque for a controllable pitch propeller in feathering maneuver. *Brodogradnja* **2023**, *74*, 74205. [CrossRef]
21. Ortolani, F.; Capone, A.; Dubbioso, G.; Pereira, F.A.; Maiocchi, A.; Felice, F.D. Propeller performance on a model ship in straight and steady drift motions from single blade loads and flow field measurements. *Ocean Eng.* **2020**, *197*, 106881. [CrossRef]
22. Grlj, C.G.; Degiuli, N.; Farkas, A.; Martić, I. Numerical Study of Scale Effects on Open Water Propeller Performance. *J. Mar. Sci. Eng.* **2022**, *10*, 1132. [CrossRef]
23. Zhang, Z.; Sun, P.; Pan, L.; Zhao, T. On the propeller wake evolution using large eddy simulations and physics-informed space-time decomposition. *Brodogradnja* **2024**, *75*, 75102. [CrossRef]
24. Lungu, A. Numerical simulation of the cavitating KP505 propeller working in open water conditions. *IOP Conf. Ser. Mater. Sci. Eng.* **2018**, *400*, 042035. [CrossRef]

25. Tokyo 2015 CFD Workshop Organizers, 2015. Tokyo 2015 CFD Workshop Website. Available online: https://t2015.nmri.go.jp/kcs_gc.html (accessed on 15 May 2024).
26. Menter, F.R. Two-equation eddy-viscosity turbulence models for engineering applications. *AIAA J.* **1994**, *32*, 1598–1605. [CrossRef]
27. Paterson, E.; Wilson, R.; Stern, F. *General-Purpose Parallel Unsteady RANS Ship Hydrodynamics Code: CFDSHIP-IOWA*; Iowa Institute of Hydraulic Research: Iowa City, IA, USA, 2003.
28. He, T.; Feng, D.; Liu, L.; Wang, X.; Jiang, H. CFD Simulation and Experimental Study on Coupled Motion Response of Ship with Tank in Beam Waves. *J. Mar. Sci. Eng.* **2022**, *10*, 113. [CrossRef]
29. Dong, K.; Wang, X.; Zhang, D.; Liu, L.; Feng, D. CFD Research on the Hydrodynamic Performance of Submarine Sailing near the Free Surface with Long-Crested Waves. *J. Mar. Sci. Eng.* **2022**, *10*, 90. [CrossRef]
30. Li, T.; Yan, C.; Li, Y. Investigation and enhancement of cut-paste algorithm in overlapping grid. *J. Beijing Univ. Aeronaut. Astronaut.* **2005**, *31*, 402–406.
31. Cho, K.W.; Kwon, J.H.; Lee, S. Development of a fully systemized chimera methodology for steady/unsteady problems. *J. Aircr.* **1999**, *36*, 973–980. [CrossRef]
32. Bonet, J.; Peraire, J. An alternating digital tree (ADT) algorithm for 3D geometric searching and intersection problems. *Int. J. Numer. Methods Eng.* **1991**, *31*, 1–17. [CrossRef]
33. Feng, D.; Yu, J.; He, R.; Zhang, Z.; Wang, X. Free running computations of KCS with different propulsion models. *Ocean Eng.* **2020**, *214*, 107563. [CrossRef]
34. Yu, J.; Yao, C.; Liu, L.; Zhang, Z.; Feng, D. Assessment of full-scale KCS free running simulation with body-force models. *Ocean Eng.* **2021**, *237*, 109570. [CrossRef]
35. AMSE. *Standard for Verification and Validation in Computational FLUID Dynamics and Heat Transfer-ASME V & V 20-2009*; The American Society of Mechanical Engineers: New York, NY, USA, 2009.
36. AMSE. *Multivariate Metrics-Supplement 1 of ASME V & V 20-2009*; The American Society of Mechanical Engineers: New York, NY, USA, 2016.
37. Feng, D.; Yu, J.; He, R.; Zhang, Z.; Wang, X. Improved body force propulsion model for ship propeller simulation. *Appl. Ocean Res.* **2020**, *104*, 102328. [CrossRef]
38. Yu, J.; Feng, D.; Liu, L.; Yao, C.; Wang, X. Assessments of propulsion models for free running surface ship turning circle simulations. *Ocean Eng.* **2022**, *250*, 110967. [CrossRef]
39. Stern, F.; Wilson, R.V.; Coleman, H.W.; Paterson, E.G. Comprehensive Approach to Verification and Validation of CFD Simulations—Part 1: Methodology and Procedures. *ASME. J. Fluids Eng.* **2001**, *123*, 793–802. [CrossRef]
40. Xing, T.; Stern, F. Factors of safety for Richardson extrapolation. *J. Fluids Eng.* **2010**, *132*, 061403. [CrossRef]
41. Tokgoz, E. A CFD Study on the Propeller-Hull Interaction Flow in Waves Using Body-Force Distribution Model. Doctoral Dissertation, Osaka University, Osaka, Japan, 2015.
42. Newman, J.N. *Marine Hydrodynamics*; MIT Press: Cambridge, MA, USA, 1977; p. 402.

Journal of
Marine Science and Engineering

MDPI

Article

Numerical Analysis and Optimization of Secondary Flow Channels in a PEMFC Cooling Plate

Wu Chen *, Yaxin Yin and Yan Liu

Merchant Marine College, Shanghai Maritime University, Shanghai 201306, China;
202330110029@stu.shmtu.edu.cn (Y.Y.); 202230110206@stu.shmtu.edu.cn (Y.L.)
* Correspondence: chenwu@shmtu.edu.cn; Tel.: +86-21-38282927

Abstract: Proton exchange membrane fuel cells (PEMFCs) offer a promising zero-emission power solution for maritime transportation, yet thermal management remains challenging due to localized overheating and non-uniform temperature distribution. To address the trade-off between pressure drop and thermal performance in marine PEMFC cooling plates, this study developed and systematically evaluated six flow channel configurations through CFD simulations. Parametric analysis coupled with orthogonal experimental design was employed to explore the effects of secondary flow channel number (N), angle (α), width (d), and spacing (L). The results demonstrated that Type B (parallel flow with secondary channels) reduced the pressure drop by 28.2% while achieving the highest cooling efficiency coefficient (2.66×10^4) compared to conventional configuration. Range analysis further ranked parameter sensitivity and identified optimal parameter combinations for distinct optimization objectives: thermal performance ($N = 7$, $\alpha = 30°$, $d = 0.5$ mm, and $L = 2.5$ mm), pressure drop ($N = 8$, $\alpha = 75°$, $d = 1.5$ mm, and $L = 2.5$ mm), and cooling efficiency ($N = 8$, $\alpha = 90°$, $d = 1.5$ mm, and $L = 2.5$ mm). These findings provide practical guidelines for designing cooling plates that address thermal-hydraulic requirements in marine PEMFC systems, advancing their viability for maritime propulsion applications.

Keywords: proton exchange membrane fuel cell (PEMFC); liquid cooling plate; secondary flow channel; thermal-hydraulic performance; orthogonal optimization

1. Introduction

Maritime transportation, accounting for approximately 80% of global trade volume [1], plays an indispensable role in the global economy. Nevertheless, traditional marine propulsion systems heavily rely on fossil fuels, contributing significantly to carbon emissions and global warming. In response, the International Maritime Organization (IMO) implemented new specifications in 2024 aiming to substantially reduce carbon emissions from international shipping [2], making the development and application of clean, efficient, and sustainable marine propulsion technology an urgent necessity for the maritime industry.

Currently, various alternative energy technologies, such as fuel cells, lithium batteries, solar photovoltaic systems, wind propulsion, and biofuels [3], are being explored and implemented in marine vessels. Among these, proton exchange membrane fuel cells (PEMFCs) have emerged as promising candidates due to their high power density, excellent conversion efficiency, rapid start-up, low operating temperature, and zero-emission [4–6]. Maritime application of fuel cell technology began in 2003 with Norway's offshore supply vessel "Viking Lady", marking a significant milestone for marine fuel cell propulsion. Since then, extensive research efforts have advanced PEMFC technology from conceptual studies to practical maritime applications.

Despite their advantages, PEMFC systems generate substantial heat during operation, presenting significant thermal management challenges, such as localized overheating and non-uniform temperature distribution. Localized overheating accelerates component degradation and increases safety risks, whereas non-uniform temperature distribution creates local hotspots, impeding proton transport and conductivity, thus reducing cell durability and stability [7]. Therefore, effective thermal management is crucial for reliable PEMFC operation in maritime environments. Various cooling techniques, including air cooling, liquid cooling, heat sink cooling, and phase change cooling, have been investigated for PEMFC thermal management [8]. While both air and liquid cooling are common methods, large-scale PEMFC stacks typically employ liquid cooling due to their superior cooling capacity compared to air cooling. Cooling plates with internal cooling flow channels effectively dissipate heat, prevent overheating, and promote uniform temperature distribution within PEMFC stacks [9,10].

The thermal-hydraulic performance of PEMFC cooling plates significantly depends on the cooling flow channel configuration [11]. Numerous innovative designs, such as biomimetic leaf-vein branch channels [12], fin channels [13], and tree-shaped fractal channels [14], have been proposed to enhance cooling efficiency. However, these complex structures often introduce manufacturing difficulties and increased costs. Comparatively, conventional parallel and serpentine channels, characterized by simple geometries, have been extensively studied and widely implemented. Many researchers have developed innovative designs based on these conventional flow channels [15,16]. Moreover, previous research has consistently shown that serpentine channels typically provide superior cooling effectiveness compared to parallel channels [17–20].

Extensive comparative studies have been conducted between conventional serpentine and other innovative serpentine channel configurations, continuously improving serpentine channel designs. Innovative variants, such as multi-pass serpentine [21,22], serpentine mini-channel heat sinks (SMCHSs) [23], and mixed serpentine (V-I) [24] demonstrate the superior performance of maximum surface temperature and temperature uniformity. To further optimize the thermal performance of serpentine channels, Deng et al. [25] investigated the effects of cooling channel number, the layout of channels, and the inlet temperature of coolant on the cooling performance of conventional serpentine variants through numerical simulation. Similarly, Hamrang et al. [26] analyzed the effects of geometric parameters, including the ratio of channel height to channel width and the ratio of rib width to channel width, on common parallel–serpentine and baffled parallel–serpentine flow channels. Researchers have also focused on the layout of the inlet and the outlet. Jaffal et al. [27] proposed a serpentine flow channel cooling plate with ribs and investigated the effects of rib shape and angle on the thermal performance of the system. The results revealed that ribs with 45° and triangles had the best thermal performance. González-Morán [28] systematically investigated the thermal performance of serpentine flow channel cooling plates, respectively, with four and six channels, to examine the functional relationship between the Nusselt number and both Reynolds and Graetz numbers. Sheng et al. [29] developed a novel serpentine channel cooling plate with dual inlets and outlets to further improve temperature uniformity. Although these designs significantly improve thermal performance, they often lead to increased pressure drop and higher pumping power requirements [30]. Addressing this issue, recent studies have explored hybrid channel designs combining parallel and serpentine structures [31], as well as serpentine channels with secondary flow channels or elliptical grooves [32], achieving substantial reductions in pressure drop and pump power consumption without compromising thermal performance significantly. At the same time, studies have summarized and compared the research on serpentine flow channels, as shown in Table 1.

Table 1. Summary of reviews on serpentine flow channels.

Reference	Subject	Remarks
Yu et al. [21]	Conventional serpentine, multi-pass serpentine, and spiral flow channels	The superior performance of multi-pass serpentine flow fields in terms of maximum surface temperature and temperature uniformity.
Baek et al. [22]	Six cooling channel designs for large-scale cooling plates	Multi-pass serpentine channels achieved better temperature distribution uniformity than conventional serpentine channels while maintaining consistent coolant pressure drops.
Imran et al. [23]	Serpentine mini-channel heat sinks (SMCHS)	Superior performance of the SMCHSs than conventional serpentine.
Wang et al. [24]	Six serpentine flow channels	The mixed serpentine (V-I) cooling plates showed the most consistent temperature distribution.
Deng et al. [25]	Conventional serpentine flow channels	The effects of cooling channel number, channel layout, and the coolant inlet temperature on the cooling performance were investigated.
Hamrang et al. [26]	Common parallel–serpentine and baffled parallel–serpentine flow channels	The effects of geometrical parameters including the ratio of channel height to channel width and the ratio of rib width to channel width were analyzed.
Jaffal et al. [27]	Serpentine flow channel with ribs	The effects of rib shape and angle on the thermal performance were evaluated.
González-Morán et al. [28]	Serpentine with different channels	Examine the functional relationship between the Nusselt number and both Reynolds and Graetz numbers
Sheng et al. [29]	Serpentine channel with dual inlets and outlets	The locations of the inlet and outlet as well as flow directions improved the temperature uniformity.
Atyabi et al. [30]	Serpentine channel with multiple parallel arrangements	Novel serpentine achieved effective heat dissipation, but with increased system pressure drop.
Sanmto et al. [31]	Hybrid parallel–serpentine oblique-fins cooling channel	Novel serpentine demonstrated reduced flow resistance and lower pump power consumption.
Fan et al. [32]	Serpentine with both elliptical grooves and secondary channels	Addressed the high-pressure drop and power consumption issues of conventional serpentine designs.

In summary, several key research gaps in the field of PEMFC cooling plate design for maritime applications are as follows: (1) Traditional serpentine channel cooling plates exhibit significant limitations in achieving optimal thermal-hydraulic performance, particularly in balancing heat dissipation with minimized pumping power requirements critical for marine propulsion efficiency. (2) The potential of secondary flow channel configurations as an enhancement strategy for maritime PEMFC cooling remains largely unexplored, with insufficient understanding of geometric parameter effects. To address these research gaps, the key contributions of this study are as follows: (1) Development and systematic evaluation of six specifically designed cooling plate configurations, comprising three baseline designs (parallel, serpentine, and hybrid channels) and three enhanced designs incorporating secondary flow channels. (2) Comprehensive parametric analysis investigating the influence of four geometric parameters characterizing secondary flow channels on thermal-hydraulic performance: the number of channels (N), channel angle (α), channel width (d), and channel spacing (L). (3) Implementation of a four-factor five-level orthogonal experimental design coupled with range analysis to quantitatively rank parameter sensitivity and identify optimal parameter combinations. These contributions significantly advance the understanding of cooling plate design for maritime PEMFCs and offer practical solutions to the thermal management challenges that currently limit their widespread adoption in marine propulsion systems.

The remainder of this paper is organized as follows: Section 2 presents the numerical simulation model, detailing the geometric and mathematical model and validation methodology employed in this study. Section 3 introduces the evaluation parameters that serve as quantitative metrics for assessing the thermal-hydraulic performance of various cooling plate designs. Section 4 provides comprehensive result analysis. Finally, Section 5 summarizes the key findings, and suggests directions for future research.

2. Numerical Simulation Model

2.1. Geometric Model

PEMFCs are typically deployed in stacked configurations, where multiple unit cells are connected in series to meet varying power requirements, ranging from several watts to megawatts. The work principle of PEMFCs involves hydrogen oxidation reaction (HOR) and oxygen reduction reaction (ORR) [33]. Each unit cell comprises bipolar plates, membrane electrode assemblies (MEAs), and sealing components. The cooling plate, positioned between two adjacent unit cells for thermal management, interfaces with the cathode side of one cell and the anode side of another [34].

To evaluate the performance of different flow channel configurations, six geometric models were developed, as depicted in Figure 1. Types A, C, and E represent conventional configurations featuring parallel, serpentine, and hybrid (parallel–serpentine combination) channels, whereas Types B, D, and F represent novel configurations, incorporating identical secondary flow channel structures into these three conventional configurations, respectively. All models were designed in SolidWorks 2021 and simulated using COMSOL Multiphysics 6.1. The cooling plate model, as depicted in Figure 2, has dimensions of 50 mm (length) $\times$ 50 mm (width) $\times$ 2 mm (thickness), with an inlet cross-sectional area of 1 mm $\times$ 1 mm. Coolant enters through the inlet, removes heat generated by the cells, and exits through the outlet. The primary channels shown in Figure 3a were designed with uniform spacing and a width of 1 mm. Secondary flow channels, introduced as auxiliary structures on the primary channels, are illustrated in Figure 3b, with consistent geometric parameters: number (N = 10), angle (α = 45°), width (d = 1 mm), and spacing (L = 2 mm).

2.2. Mathematical Model

2.2.1. Heat Generation Rate Model

Currently, Larminie et al. [35] employ theoretical analysis to calculate the heat generation rate in PEMFCs. Assuming uniform heat generation within the cell and water product existing in vapor form, the heat generation and heat flux can be calculated using Equations (1) and (2):

$$Q = I\left(-\frac{\Delta h_f^0}{2F} - V\right) \tag{1}$$

$$q = \frac{Q}{2A} \tag{2}$$

where I and V represent the output current and voltage of a unit cell, respectively; Δh_f^0 denotes the lower heating value of water (-237.3 kJ·mol^{-1}); F is the Faraday constant (96,487 C·mol^{-1}); and A represents the surface area of one side of the cooling plate.

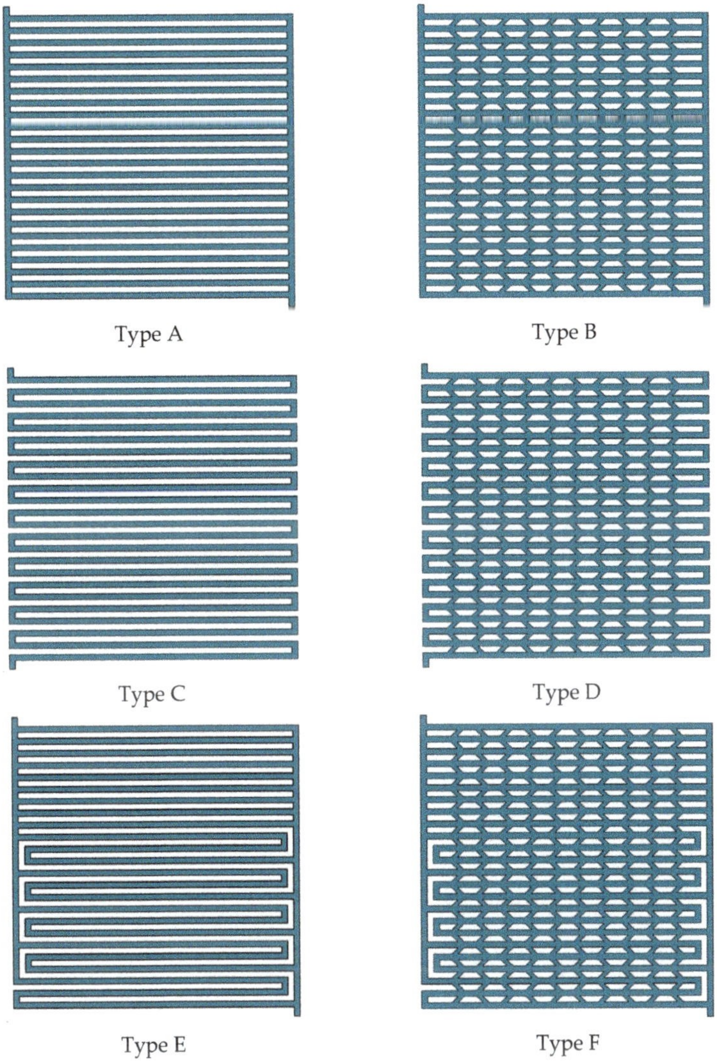

Figure 1. Configurations of six flow channel designs.

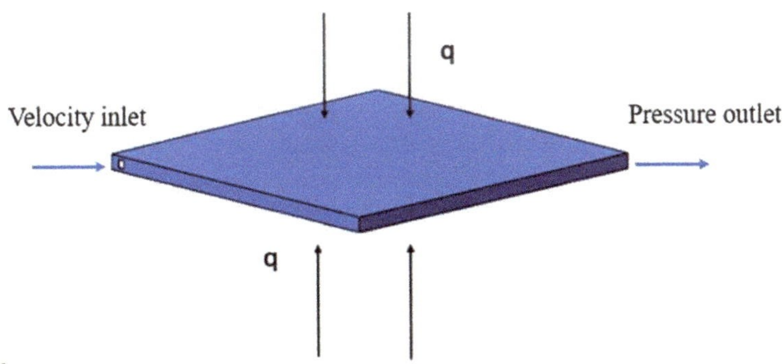

Figure 2. Schematic diagram of the liquid cooling plate.

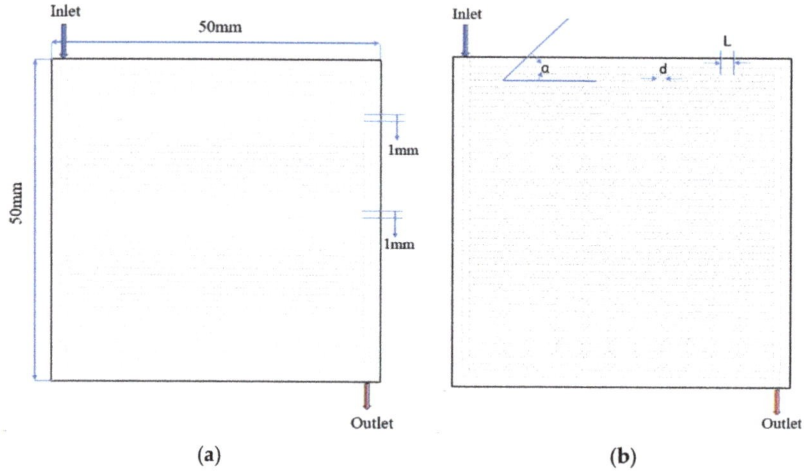

Figure 3. Geometric diagram of (**a**) conventional and (**b**) novel flow channel structures.

2.2.2. Governing Equation

The three-dimensional fluid dynamics within PEMFC cooling plates were numerically simulated using COMSOL Multiphysics. Specifically, the CFD Module and Heat Transfer Module were employed to solve the coupled fluid flow and heat transfer problem. Assuming the material of cooling plates is homogeneous and isotropic, the solid domain governing equation is given by Equation (3):

$$\frac{\partial}{\partial x_j}\left(k_g \frac{\partial T}{\partial x_j}\right) = 0 \tag{3}$$

where k_g represents the thermal conductivity of the solid material.

2.2.3. Boundary Conditions and Physical Parameters

In order to simplify the calculation, the following assumptions are made in this study:

(1) The coolant is treated as an incompressible Newtonian fluid with constant density.
(2) The cooling plate material is considered homogeneous and isotropic, with temperature-independent physical parameters.
(3) At the inlet of cooling plates, the flow distribution is uniform. The no-slip condition is imposed at fluid–wall interfaces, ensuring zero relative velocity between the coolant and channel walls.
(4) The temperature between the solid and fluid domains is coupled, allowing conjugate heat transfer.
(5) Gravitational and buoyancy effects are neglected throughout the computational domain.

Based on the geometric model, the computational domain consists of two regions: the solid domain (cooling plate) and the fluid domain (coolant). The cooling plate is made of graphite. Various coolants such as water, dielectric fluids, and ethylene glycol have been employed in thermal management systems. Among these options, water remains the most widely utilized due to its cost-effectiveness and accessibility. Despite water's relatively lower thermal conductivity posing certain limitations for battery thermal management applications [36], research-trending Al_2O_3 nanofluids have not gained widespread adoption, primarily due to their prohibitive cost. As shown in Table 2, water exhibits higher thermal conductivity and specific heat capacity compared to 50% ethylene glycol solution, resulting in superior heat transfer performance while substantially reducing cooling costs. Therefore, water was selected as the coolant medium for this study. The physical properties parameters are shown in Table 3.

Table 2. Comparison of thermal properties between water and 50% ethylene glycol solution.

Coolant	Density ($kg \cdot m^{-3}$)	Specific Heat Capacity ($J \cdot kg^{-1} \cdot K^{-1}$)	Thermal Conductivity ($W \cdot m^{-1} \cdot K^{-1}$)
50% Ethylene glycol solution	1071.1	3300	0.384
Water	983.1	4179	0.659

Table 3. Physical properties of materials.

Parameters (Unit)	Value
Density of graphite cooling plate ($kg \cdot m^{-3}$)	2250
Specific heat capacity of graphite cooling plate ($J \cdot kg^{-1} \cdot K^{-1}$)	691
Thermal conductivity of graphite cooling plate ($W \cdot m^{-1} \cdot K^{-1}$)	85.5
Density of water ($kg \cdot m^{-3}$)	983.1
Specific heat capacity of water ($J \cdot kg^{-1} \cdot K^{-1}$)	4179
Thermal conductivity of water ($W \cdot m^{-1} \cdot K^{-1}$)	0.659
Dynamic viscosity of water ($Pa \cdot s$)	0.00047

The heat flux generated by PEMFC exhibits a proportional relationship with current density across operational voltage conditions. Although the actual heat flux distribution across cooling plates is inherently non-uniform, Afshari et al. [37] proposed that assuming uniform and constant heat flux provides a reasonable approximation of cooling plate performance under real operation conditions. For numerical implementation, constant heat flux boundary conditions were imposed on both the upper and lower surfaces of the cooling plate, with the cooling plate thermal performance serving as a proxy for PEMFC thermal performance. The remaining four surfaces were treated as adiabatic boundaries.

To determine the heat flux applied to the cooling plate, heat power generation performance tests were conducted on a fuel cell unit. The schematic diagram of the test system is shown in Figure 4, and the experimental setup is shown in Figure 5. The corresponding heat flux was calculated from the measured voltage values at different current densities using the heat generation rate model. Before testing, nitrogen gas was purged through the test station at 2 $L \cdot min^{-1}$ for 300 s to remove residual air contaminants. After purging, the fuel cell was heated to an operating temperature of 85 °C using heating films. Simultaneously, air and hydrogen (100% relative humidity) were supplied to the cathode and anode sides, respectively, with stoichiometric flow ratios of 1.95:1.5. The gas flow rates were automatically regulated by the test station based on these stoichiometric flow ratios, maintaining an operating pressure of 101.325 kPa. The fuel cell had an active area of 25 cm^2, comprising cathode and anode bipolar plates, MEA, current collectors, sealing components, and fastening devices, as shown in Figure 6. The bipolar plates were made of graphite, and a Nafion-117 membrane was used, secured by custom-designed fixtures. During experimentation, detailed measurements were conducted in the activation region (low-current-density region). The current density was incrementally increased from 0 to 2 $A \cdot cm^{-2}$ in steps of 0.2 $A \cdot cm^{-2}$. The polarization curves were obtained using a Scribner 850 Fuel Cell Test System employing the current scan method. To ensure steady-state conditions, voltage measurements were recorded after maintaining each current density for approximately 5 min, with final values averaged over the last minute of each measurement period. After testing, nitrogen purging was performed on the hydrogen lines to ensure complete hydrogen removal, and the system was only shut down after completely releasing the pressure from all gas lines.

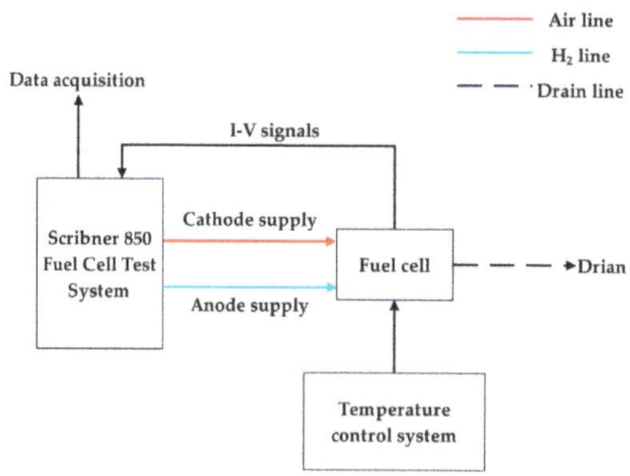

Figure 4. Schematic diagram of test system.

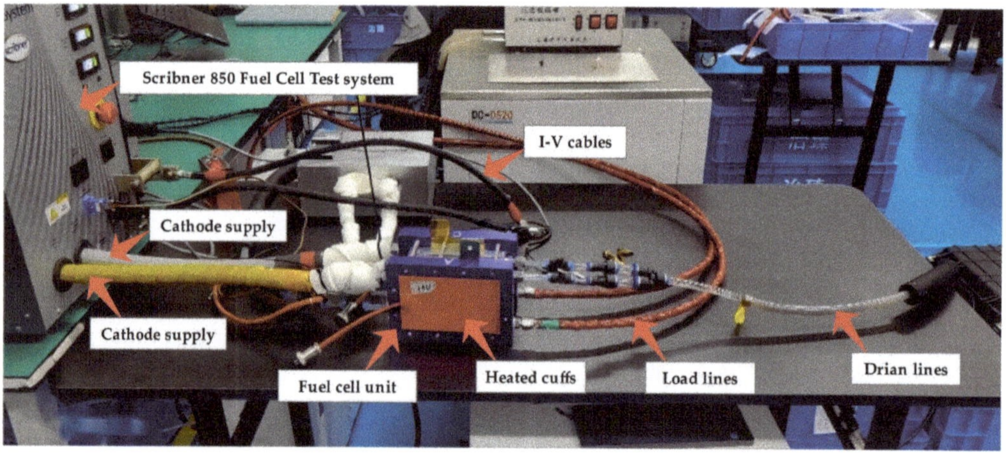

Figure 5. The fuel cell test station.

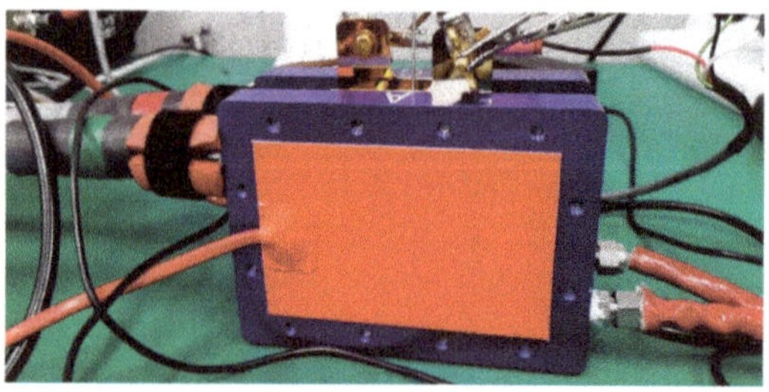

Figure 6. The test fuel cell unit.

Due to measurement uncertainty associated with measuring equipment, it is essential to evaluate their impact on the results. The parameters in this study can be categorized into directly measured parameters and derived parameters. The directly measured parameters include the operating current density and output voltage of the fuel cell, while the derived parameter is the heat flux of fuel cell. To assess the impact of these uncertainties on experimental data processing, an uncertainty analysis was conducted according to

Equations (4) and (5) [38]. Table 4 summarizes the accuracy of measuring equipment and the uncertainty of the main parameters.

$$\Delta Y = \sqrt{\sum_i \left(\frac{\partial Y}{\partial X_i}\right)^2 \Delta X_i^2} \tag{4}$$

$$R_Y = \Delta Y / Y \tag{5}$$

where $\Delta Y/Y$ is the sensitivity coefficient of the parameter Y relative to X_i, ΔX_i represents the uncertainty of X_i, ΔY is the uncertainty of Y, and R_Y is the relative uncertainty of paramotor Y.

Table 4. The measuring equipment accuracy and the main parameters' relative uncertainty.

Parameters	Range	Accuracy	Unit
Current density	0–10	±0.3%FS	A·cm^{-2}
Output voltage	0.2–20	±0.3%	V
Heat flux	—	±0.897%	W·cm^{-2}

Based on the PEMFC heat generation rate model, the test data were substituted to calculate the corresponding heat flux under different operating conditions. Figure 7 shows the measured polarization curve and the variation in heat flux with current density. Assuming 80% of the fuel cell heat was removed by coolant, four current densities were selected: 0.6, 1.0, 1.4, and 1.8 A·cm^{-2}, corresponding to the heat flux of 0.107, 0.195, 0.296, and 0.404 W·cm^{-2}, respectively. These values were used as thermal boundary conditions for subsequent analysis of the thermal performance of different cooling plates.

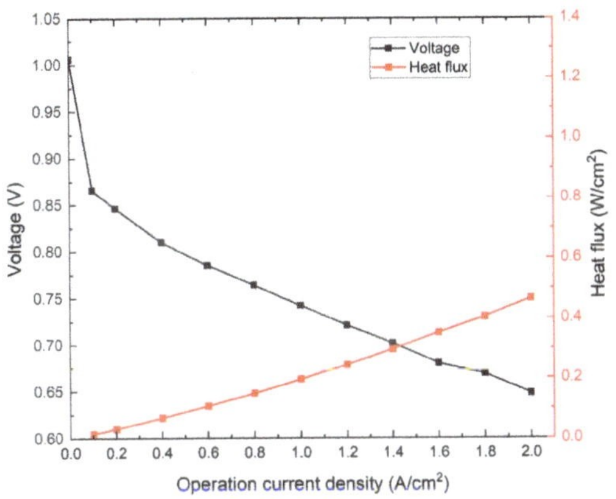

Figure 7. Diagram of polarization curve and variation in heat flux with current density.

When investigating the effect of coolant inlet velocity on the performance of cooling plates, a heat flux of 0.25 W·cm^{-2} (corresponding to a current density of 1.2 A·cm^{-2} from the above tests) was imposed on both the upper and lower surfaces of the cooling plate. A velocity inlet boundary condition was applied with the flow direction perpendicular to the inlet face. Five inlet velocities were selected (0.3, 0.4, 0.5, 0.6, and 0.7 m·s^{-1}) for CFD simulation. The boundary condition at the outlet was set as a pressure outlet with a value of 0 Pa.

For fluid flow in channels, the Reynolds number (*Re*) is commonly employed to characterize flow regimes. When $Re \leq 2300$, the flow is in a laminar state; when $2300 \leq Re \leq 4000$, the flow is in a transitional state; and when $Re \geq 4000$, the flow is in a turbulent state. The Reynolds number is calculated as shown in Equation (6):

$$Re = \frac{\rho u_{in} d}{\mu} \quad (6)$$

where d represents the characteristic length and μ is the dynamic viscosity of the coolant. In this study, the flow channels are rectangular. Based on calculations, the Reynolds numbers corresponding to the five flow velocity conditions were all less than 2300, confirming that the coolant flow in this study was in the laminar regime.

2.2.4. Grid Independence Verification

Mesh generation is fundamental to numerical simulation, with mesh quantity and quality determining both computational accuracy and time consumption. In this study, grid independence verification was performed on the Type A cooling plate model using physics-controlled meshing in COMSOL Multiphysics. Five mesh densities were investigated: extra-coarse (150,000 elements), coarse (300,000 elements), normal (480,000 elements), fine (1.29 million elements), and extra-fine (2.4 million elements). Figure 8 illustrates the variations in surface average temperature and inlet–outlet pressure drop under different mesh densities. The results indicate that both the average temperature and pressure drop decrease with increasing mesh density. When the number of mesh elements exceeds 480,000, the variation in surface average temperature is less than 0.04%, and the pressure drop variation rate remains below 3%, indicating that further mesh refinement has minimal impact on computational results. Therefore, to optimize computational efficiency and resource utilization, the mesh model with 480,000 elements was selected for subsequent numerical simulations.

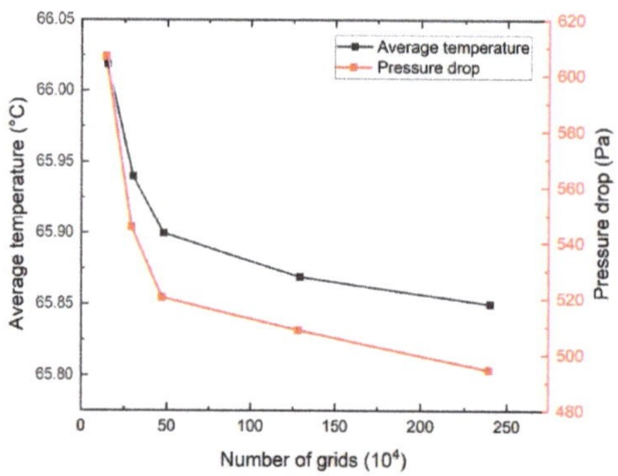

Figure 8. Grid independence verification for the Type A cooling plate.

2.2.5. Model Validation

To ensure the accuracy of subsequent simulations, the numerical model was validated against the experimental data from Yu et al. [21]. Model A from their study was selected as the validation case, with geometric dimensions, material physical properties, and boundary conditions exactly matching those specified in their work. As shown in Figure 9, the simulation results for both the maximum temperature difference and the inlet–outlet pressure drop demonstrate good agreement with the experimental measurements at varying coolant inlet velocities. The maximum deviation between simulation and experimental results was 7.4%, confirming the high accuracy of the present numerical model in predicting the thermal-hydraulic performance of cooling plates.

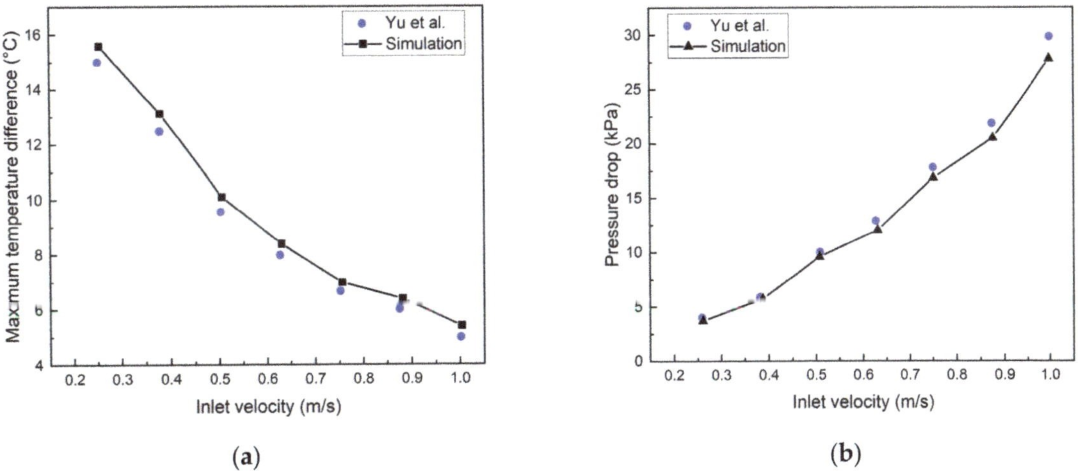

Figure 9. Comparison of simulation and experimental values of (**a**) maximum temperature difference and (**b**) pressure drop [21].

3. Evaluation Parameters

The evaluation parameters defined in this study to evaluate the cooling plate include the maximum surface temperature, maximum temperature difference, temperature uniformity index, inlet–outlet pressure drop, and cooling efficiency coefficient. To evaluate the temperature uniformity of the cooling plate, the temperature uniformity index (TUI) is introduced as one of the evaluation parameters. A smaller temperature uniformity index indicates more uniform temperature distribution on the cooling plate surface, which is expressed as Equation (7):

$$U_T = \frac{\int A|T - T_{avg}|dA}{\int AdA} \tag{7}$$

where A is the bottom surface area of the cooling plate and T_{avg} is the average temperature of the cooling plate, defined by Equation (8):

$$T_{avg} = \frac{\int ATdA}{\int AdA} \tag{8}$$

To comprehensively evaluate the cooling plate's performance, Fan et al. [32] have employed evaluation parameters to assess its heat dissipation capacity and power consumption. The ratio between these two factors defines the cooling efficiency coefficient, which reflects the effectiveness of heat transfer per unit pump work, as expressed in Equation (9):

$$j_w = \frac{Q_w}{P_w} \tag{9}$$

where Q_w is the heat removed by the coolant and P_w is the pump work consumed at the coolant inlet and outlet, defined by Equations (10) and (11):

$$Q_w = m_{in}c_p(T_{out} - T_{in}) \tag{10}$$

$$P_w = u_{in} \times A_{in} \times \Delta P \tag{11}$$

Substituting Equations (10) and (11) into Equation (9), the cooling efficiency coefficient can be simplified to Equation (12):

$$j_w = \frac{\rho c_p (T_{out} - T_{in})}{\Delta P} \tag{12}$$

where m_{in} is the mass flow rate of the inlet coolant. A larger j_w indicates better comprehensive performance of the cooling plate.

4. Result Analysis and Discussion

Through numerical simulations, the performance of the cooling plate was evaluated using the above evaluation parameters. The effects of heat flux, coolant inlet velocity, and secondary flow channel structural parameters on cooling plate performance were systematically analyzed. Subsequently, an orthogonal experimental design with four factors at five levels was established. Range analysis was employed to investigate the sensitivity of these four factors under different optimization objectives when acting simultaneously. Finally, optimal combinations of secondary flow channel structural parameters were proposed for different optimization objectives.

4.1. Effect of Heat Flux on the Performance of Cooling Plates

This section analyzes the heat dissipation performance of six cooling plate designs under four different heat flux conditions (0.107, 0.195, 0.296, and 0.404 W·cm^{-2}) with a coolant inlet velocity of 0.6 m·s^{-1}. As shown in Figure 10, under the same heat flux, the Type A cooling plate exhibits the highest maximum surface temperature and maximum temperature difference, followed by Type D, while Type C shows the lowest values. At a heat flux of 0.107 W·cm^{-2}, the maximum surface temperatures of the six cooling plates, in descending order, are as follows: Type A (63.81 °C), Type D (63.74 °C), Type B (63.49 °C), Type F (63.30 °C), Type E (62.78 °C), and Type C (62.72 °C). The maximum temperature differences among the six designs are relatively small at a low heat flux, with variations of approximately 1 °C. The trend in maximum temperature difference mirrors that of the maximum surface temperature. At 0.107 W·cm^{-2}, the difference in maximum temperature variation between Type A and Type C cooling plates is 1.1 °C. When the heat flux increases to 0.404 W·cm^{-2}, Type A still maintains the highest maximum surface temperature and temperature difference, with increases of 10.59 °C and 10.65 °C, respectively. However, Type B, which incorporates secondary flow channels based on Type A, shows smaller increases of 9.02 °C and 9.07 °C, respectively.

Furthermore, as heat flux increases, both the average temperature and temperature uniformity index of all six designs increase, indicating that higher heat flux compromises the uniform temperature distribution across the cooling plate surface. Under the same heat flux, Type A exhibits the highest temperature uniformity index, while Type E shows the lowest. At 0.107 W·cm^{-2}, the temperature uniformity index for Type A is 0.81 and 0.49 for Type E, with other designs showing similar values. When heat flux density increases to 0.404 W·cm^{-2}, Type A's temperature uniformity index increases by 2.26, while Types B (2.10), F (1.95), and D (1.86) demonstrate similar temperature uniformity performance. Types C (1.58) and E (1.35) show the smallest increase in temperature uniformity index, indicating superior temperature uniformity performance under varying heat flux.

Comprehensive analysis reveals that the Type C cooling plate demonstrates superior heat dissipation performance with increasing heat flux, attributed to its secondary flow channel structures. These structures generate vortices that enhance mixing and contact between the coolant and cooling plate, thereby promoting heat transfer. However, as heat flux increases, only parallel flow channels with secondary flow structures show improved thermal performance, while the other two traditional channel designs with additional structures do not exhibit enhanced thermal performance.

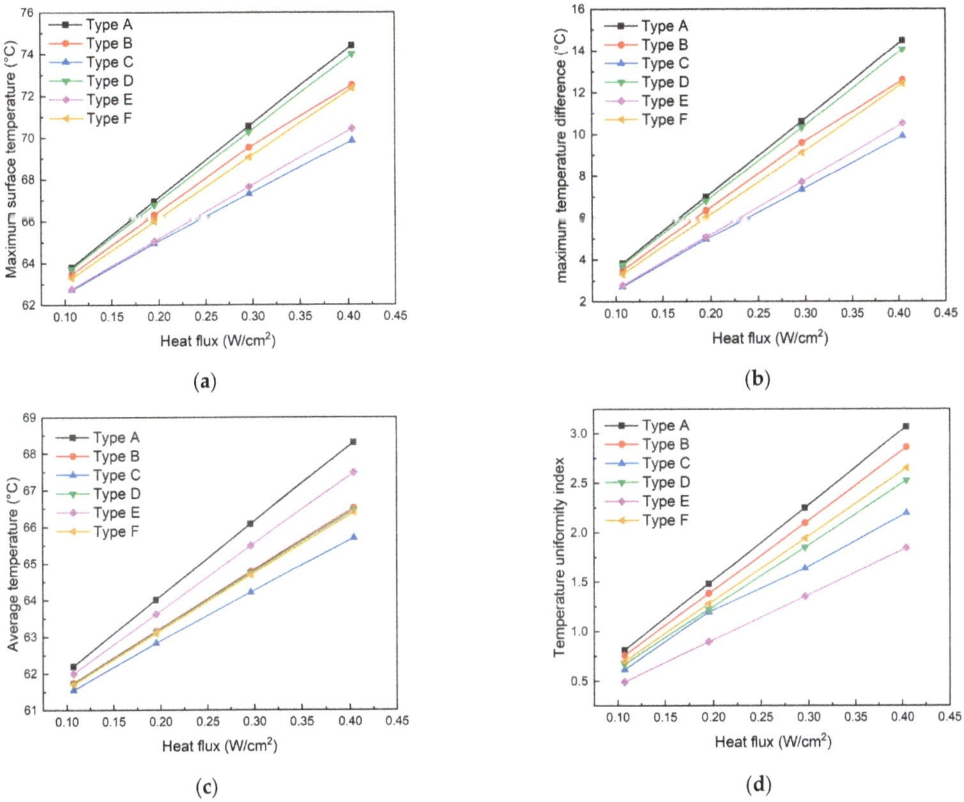

Figure 10. Effect of heat flux on (**a**) maximum surface temperature, (**b**) maximum temperature difference, (**c**) average temperature, and (**d**) TUI of six cooling plates.

4.2. Effect of Coolant Inlet Velocity on the Thermal-Hydraulic Performance of Cooling Plates

4.2.1. Effect of Coolant Inlet Velocity on the Thermal Performance of Cooling Plates

Figure 11 illustrates the surface temperature distribution contours of six cooling plate configurations at an inlet flow velocity of 0.5 $m{\cdot}s^{-1}$. The coolant inlet is positioned in the upper-left corner of each cooling plate, with the outlet located in the lower-right corner. As coolant flows through the channels, it effectively removes excess heat, causing its temperature to increase progressively while its heat transfer capacity diminishes. Consequently, a temperature gradient develops along the flow direction toward the outlet.

Notably, Type A and Type E cooling plates exhibit distinct local hotspots on their right sides, resulting in non-uniform temperature distributions. In contrast, Type B and Type F plates, which incorporate secondary flow channels, demonstrate significant mitigation of these localized hotspots. The serpentine structures of Type C and Type D channels feature multiple turns that enhance heat transfer during fluid directional changes, resulting in superior temperature uniformity. However, this improved thermal performance implies that these configurations will likely experience higher inlet–outlet pressure drops due to their complex flow paths.

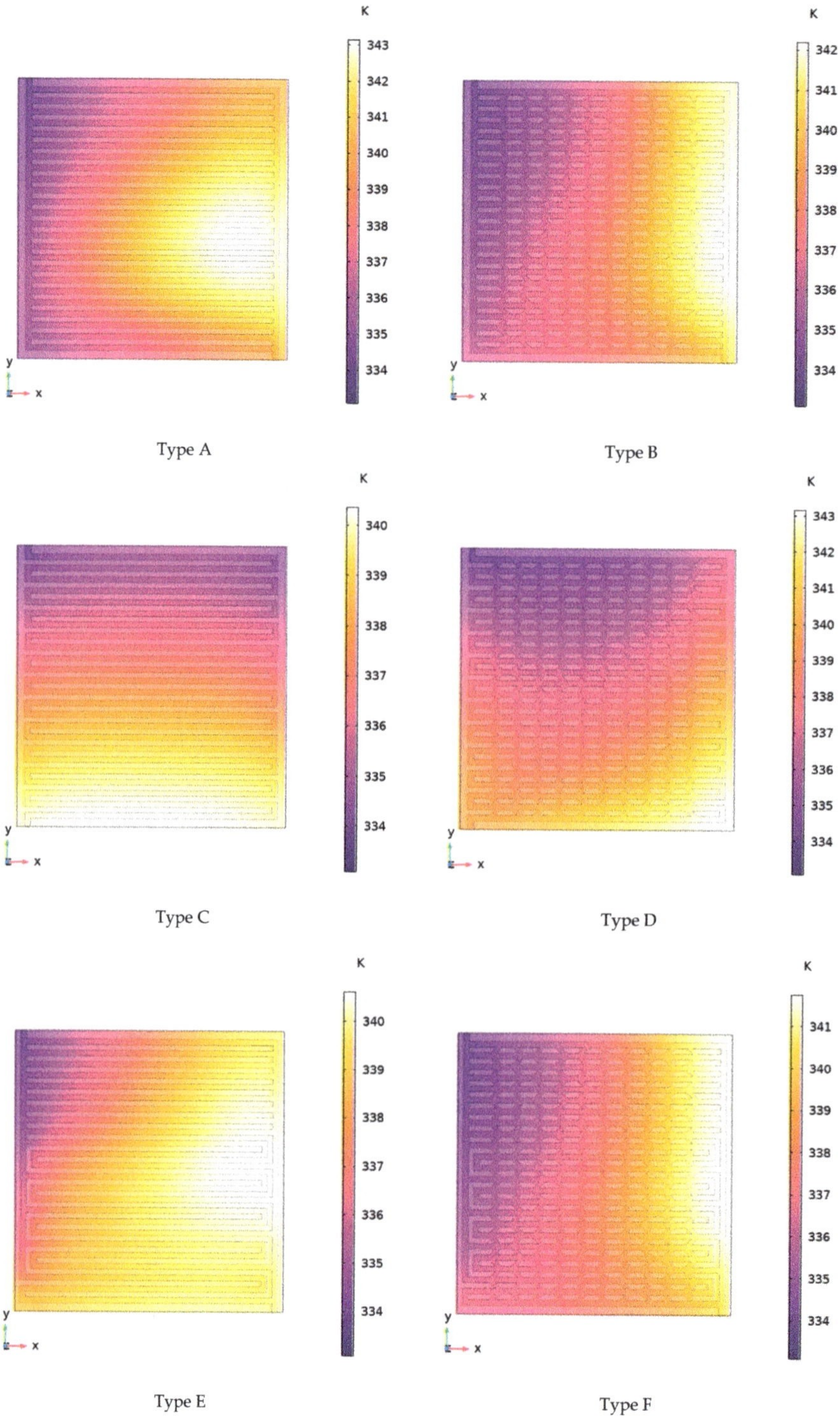

Figure 11. Temperature distribution contours of six cooling plate configurations at an inlet flow velocity of 0.5 m·s^{-1}.

As shown in Figure 12, the heat dissipation performance of all six cooling plates improves with increasing coolant inlet velocity, which is attributed to enhanced heat exchange capacity at higher flow rates. Before the inlet velocity reaches 0.5 m·s^{-1}, the Type D cooling plate exhibits the highest surface maximum temperature and temperature difference, reaching 75.23 °C and 14.34 °C, respectively, at 0.3 m·s^{-1}, significantly higher than other designs. Above 0.5 m·s^{-1}, the Type A cooling plate surpasses Type D, showing the highest surface temperature among all designs, with values of 68.09 °C and 7.13 °C for maximum surface temperature and temperature difference, respectively, at 0.7 m·s^{-1}. Type C consistently maintains the lowest values for both parameters. When inlet velocity increases from 0.3 m·s^{-1} to 0.7 m·s^{-1}, the Type D cooling plate demonstrates the largest reduction in surface maximum temperature and temperature difference, decreasing by 7.50 °C and 7.57 °C, respectively.

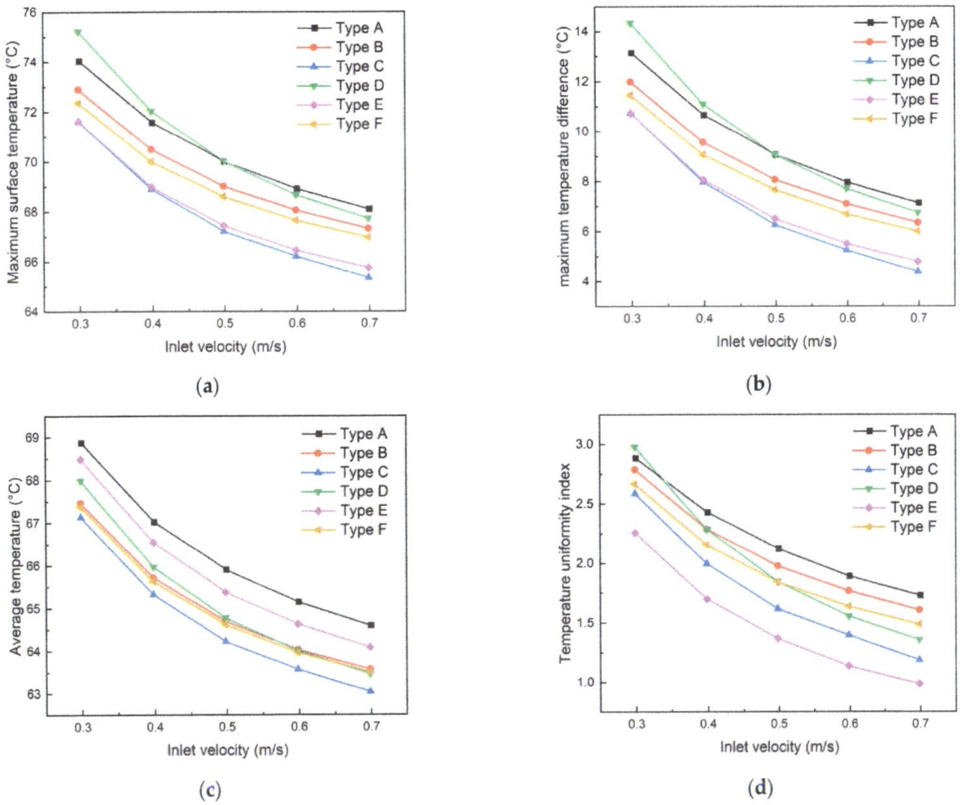

Figure 12. Effect of coolant inlet velocity on (**a**) maximum surface temperature, (**b**) maximum temperature difference, (**c**) average temperature, and (**d**) TUI of six cooling plates.

Furthermore, the average temperature of all six cooling plates gradually decreases with increasing inlet velocity, though the rate of decrease diminishes, indicating saturation in convective heat transfer effectiveness. At identical inlet velocities, Type A exhibits the highest average temperature, reaching 65.90 °C at 0.5 m·s^{-1}, followed by Type E (65.38 °C), Type D (64.78 °C), Type B (64.70 °C), Type F (64.63 °C), and Type C (64.23 °C). The consistently lower average temperature of Type C can be attributed to its serpentine channel structure, which provides multiple convective heat transfer opportunities at the bends.

Similarly, the temperature uniformity index of all designs shows a decreasing trend with increasing inlet velocity. At 0.3 m·s^{-1}, Type E demonstrates the lowest temperature uniformity index (2.26), while Type D shows the highest (2.98), with other designs ranging from high to low as follows: Type A (2.88), Type B (2.79), Type F (2.67), and Type C (2.59). However, beyond 0.4 m·s^{-1}, Type A exhibits the highest temperature uniformity index. Notably, Type D shows the most significant reduction in temperature uniformity index (54.4%) when the inlet velocity increases from 0.3 m·s^{-1} to 0.7 m·s^{-1}.

Analysis from the perspective of secondary flow channel structure reveals that only parallel channels demonstrate improved thermal characteristics (reduced maximum surface temperature, temperature difference, and temperature uniformity index) after incorporating secondary flow structures, while the other two traditional channel designs show no thermal performance enhancement.

4.2.2. Effect of Coolant Inlet Velocity on the Inlet–Outlet Pressure Drop of Cooling Plates

Figure 13 presents the pressure distribution contours of six channel configurations at an inlet flow velocity of 0.5 m·s^{-1}. In PEMFC liquid cooling systems, the power required by the coolant pump is supplied by the fuel cell itself. Consequently, increased inlet–outlet pressure drops lead to higher pump power consumption, resulting in reduced net power output from the fuel cell.

As observed in Figure 13, pressure gradually decreases along the flow direction, with maximum values at the inlet and minimum values at the outlet. The addition of secondary flow channels creates intersecting flow paths that effectively increase the coolant flow area and promote more uniform pressure distribution. Furthermore, all traditional channel designs exhibit reduced pressure drops when supplemented with secondary flow channels. This effect is particularly pronounced in serpentine channels, where Type D configuration demonstrates a pressure drop approximately one order of magnitude lower than Type C. Therefore, incorporating secondary flow channel structures contributes to pressure drop reduction and more uniform pressure distribution, ultimately achieving the objective of lower parasitic power consumption.

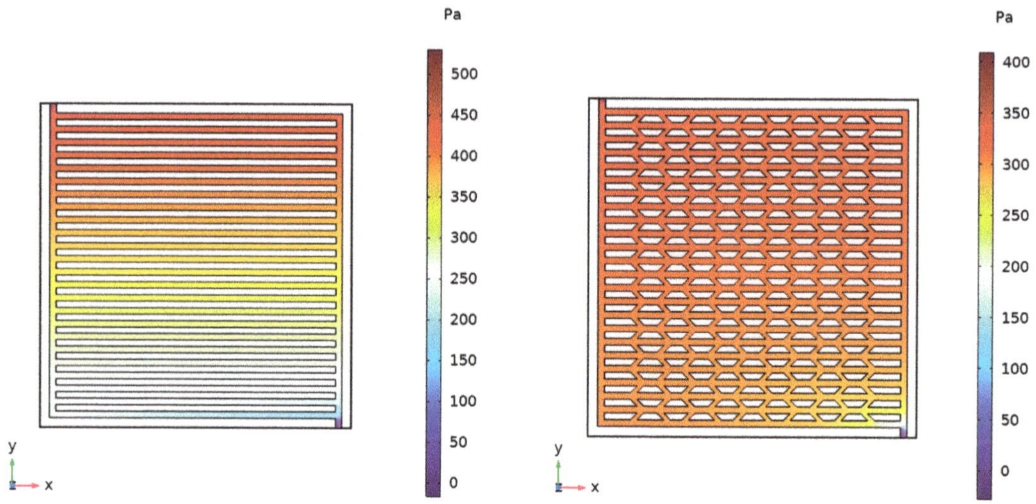

Type A

Type B

Figure 13. *Cont.*

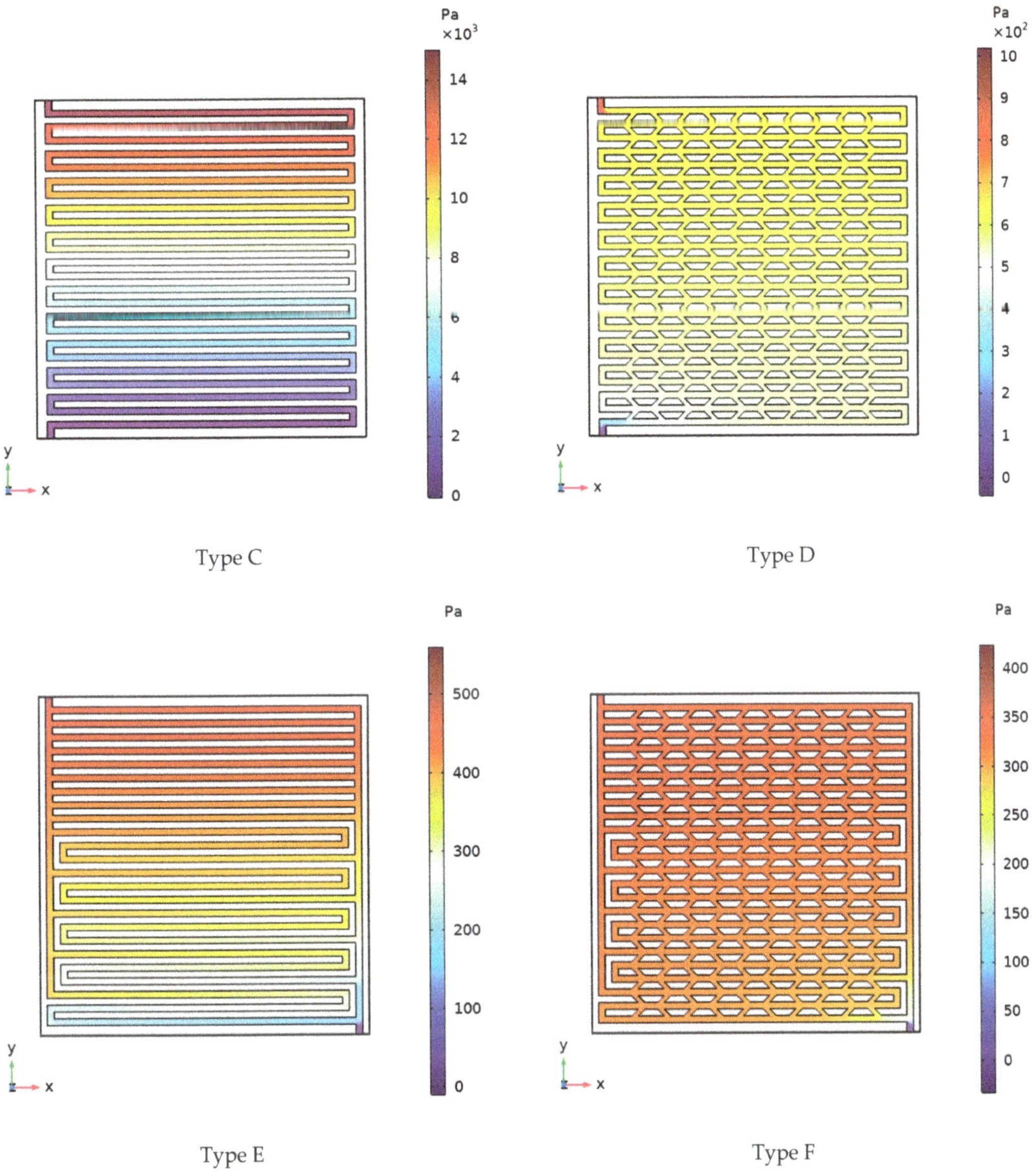

Figure 13. Pressure distribution contours of six cooling plate configurations at an inlet flow velocity of 0.5 m·s^{-1}.

Figure 14 illustrates the variation trends of inlet–outlet pressure drop across six channel designs under different coolant inlet velocities. Due to the pressure drop in the Type C channel being one to two orders of magnitude higher than the other five designs, the results are presented in two separate graphs. The analysis reveals that the inlet–outlet pressure drops of all six channels increase with increasing coolant inlet velocity. Under identical flow velocities, the Type C channel exhibits the highest pressure drop, followed by Type D; this is attributed to their serpentine structures, which extend the coolant flow path and, consequently, increase the frictional resistance. In contrast, Types B and F channels demonstrate the lowest pressure drops. At an inlet velocity of 0.3 $m{\cdot}s^{-1}$, the pressure drop in the Type C channel reaches 7068 Pa, while Types B and F channels show significantly lower values of only 171 Pa and 177 Pa, respectively. When the inlet velocity increases to 0.7 $m{\cdot}s^{-1}$, the pressure drop in the Type C channel escalates to 25,542 Pa, an increase of 18,474 Pa compared to its value of 0.3 $m{\cdot}s^{-1}$. Meanwhile, the Type B channel maintains the lowest pressure drop among all designs at 721 Pa, representing a modest increase of only 550 Pa from its value at 0.3 $m{\cdot}s^{-1}$.

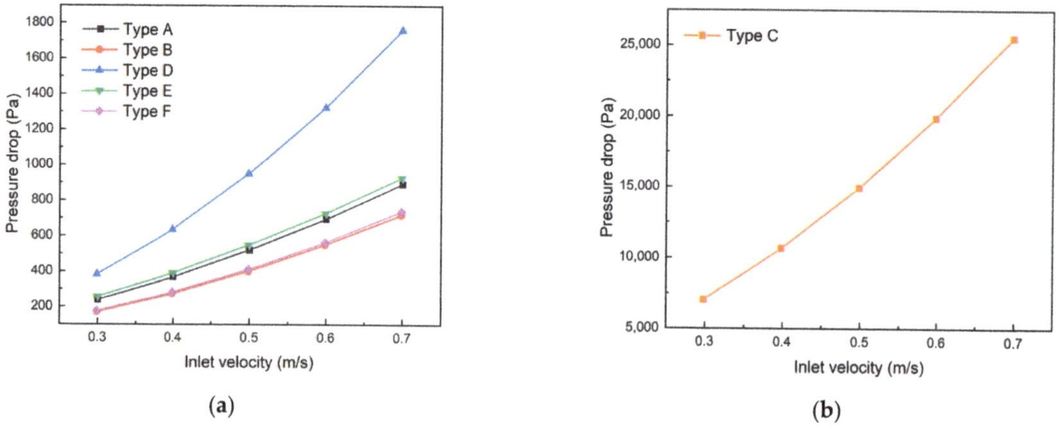

Figure 14. Effect of coolant inlet velocity on pressure drop of (**a**) Type A-F and (**b**) Type C cooling plates.

Furthermore, the incorporation of secondary flow channels significantly reduces the pressure drops in all three traditional channel designs. This effect is most pronounced at an inlet velocity of 0.3 $m{\cdot}s^{-1}$, where Type D channel shows a 94.6% reduction in pressure drop compared to Type C, followed by Type F with a 31.4% reduction compared to Type E, and Type B with a 28.3% reduction compared to Type A. The study demonstrates that the introduction of secondary flow channels effectively optimizes pressure distribution, reduces coolant flow resistance, and thereby enhances overall system performance.

4.2.3. Effect of Coolant Inlet Velocity on the Cooling Efficiency Coefficient of Cooling Plates

In this section, the cooling efficiency coefficient was employed to evaluate the heat dissipation capacity and power consumption of the cooling plates. This coefficient indicates the effectiveness of heat transfer per unit pumping power, where a higher value represents better heat transfer performance per unit pumping power.

As shown in Figure 15, the cooling efficiency coefficients of all six cooling plates decrease with increasing inlet velocity, indicating a decline in heat transfer effectiveness per unit pumping power. Although the heat removal by coolant increases, the pumping power consumption increases more significantly. In other words, the heat transfer per unit pumping power decreases, leading to reduced system energy utilization efficiency. This phenomenon is particularly evident in the Type C channel, which, despite having superior thermal characteristics compared to other designs, exhibits an inlet–outlet pressure drop one to two orders of magnitude higher than the other five channels. This results in the lowest cooling efficiency coefficient (0.08×10^4 at 0.7 $m{\cdot}s^{-1}$ inlet velocity), implying that the Type C channel requires substantially more pumping power to transfer the same amount of heat. Conversely, the Type B channel demonstrates the highest cooling efficiency coefficient (2.66×10^4 at 0.7 $m{\cdot}s^{-1}$ inlet velocity), indicating maximum heat removal per unit of pumping power.

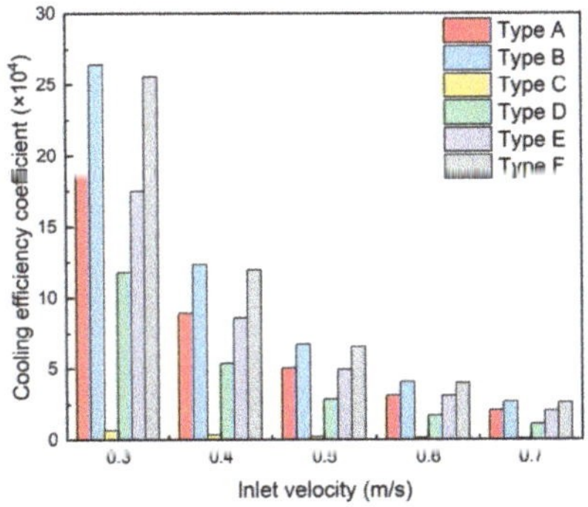

Figure 15. Comparison of cooling efficiency coefficients of six cooling plates at different inlet velocities.

Furthermore, at the same inlet velocity, all three traditional channels show increased cooling efficiency coefficients after incorporating secondary flow channels. At an inlet velocity of 0.3 m·s^{-1}, Type B channel's cooling efficiency coefficient increases by 43.4% compared to Type A, Type D's coefficient is 18.1 times that of Type C, and Type F's coefficient increases by 46.3% compared to Type E. Therefore, the addition of secondary flow channel structures effectively improves the heat transfer efficiency per unit of pumping power. Based on this comprehensive analysis, the Type B channel exhibits the best overall performance among the six cooling channel designs and is selected for subsequent secondary flow channel optimization analysis.

4.3. *Effect of Secondary Flow Channel Structural Parameters on the Performance of Cooling Plates*

Based on the secondary flow channel structure in the Type B channel, this section employs a single-factor control variable method to investigate the effects of secondary channel parameters on cooling plate heat dissipation performance and inlet–outlet pressure drop. The parameters studied include channel number (*N*), angle (α), width (*d*), and spacing (*L*).

4.3.1. Effect of the Number (N) of Secondary Flow Channels on the Thermal-Hydraulic Performance of Cooling Plates

Based on the Type B channel, five cooling plate geometric models were established with secondary channel quantities of 4, 5, 6, 7, and 8, while other parameters remained constant: channel angle at 45°, channel width at 1 mm, channel spacing at 2 mm, and heat flux at 0.25 W·cm^{-2}. All other boundary conditions and material properties remained unchanged from previous settings.

As shown in Figure 16, under the same coolant inlet velocity, the cooling plate with 8 secondary channels exhibited the lowest values across all four thermal performance indicators, although the impact of varying channel quantities on heat dissipation performance was relatively minor. At an inlet velocity of 0.5 m·s^{-1}, the eight-channel configuration demonstrated the lowest surface maximum temperature (69.21 °C) and temperature difference (8.27 °C), while the six-channel configuration showed the highest values (69.31 °C and 8.37°C, respectively), with only approximately 0.1 °C difference between them. The average temperature of the cooling plate decreased with an increase in the number of channels, but the variation was not significant. At an inlet velocity of 0.3 m·s^{-1}, the maximum temperature difference between the eight-channel and four-channel configurations was 0.25 °C, and this difference decreased with increasing flow velocity, reaching only 0.19 °C at 0.7 m·s^{-1}. As shown in Figure 16d, the number of channels had minimal impact on the temperature uniformity index. This limited effect can be attributed to the existing thermal load and coolant flow velocity conditions, which already provided relatively uniform temperature distribution, thereby limiting the potential improvements from increasing the number of channels. Overall, the influence of channel number variation on cooling plate heat dissipation performance was not substantial.

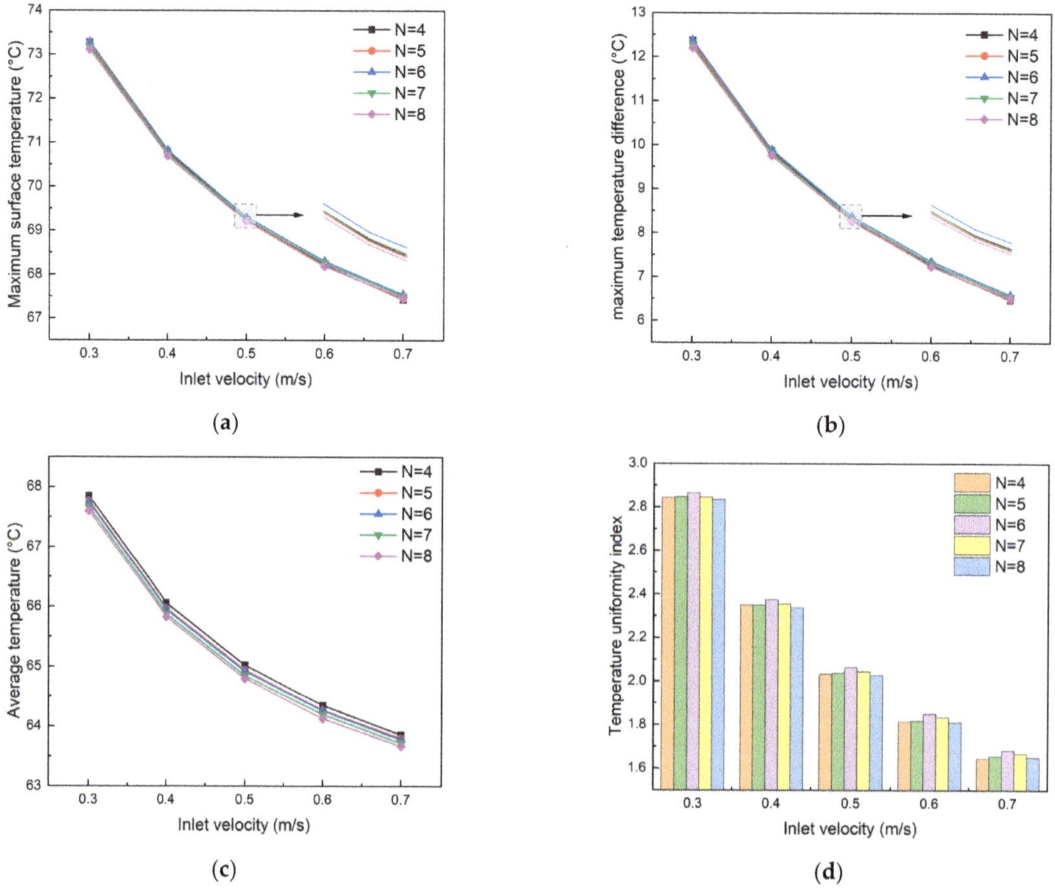

Figure 16. Effect of the number of secondary flow channels on (**a**) maximum surface temperature, (**b**) maximum temperature difference, (**c**) average temperature, and (**d**) TUI of cooling plates.

As illustrated in Figure 17, under the same flow velocity, the inlet–outlet pressure drop decreased as the number of secondary channels increased, achieving a maximum reduction of 12.7% (at 0.3 m·s^{-1} inlet velocity). This reduction occurred because the increased number of channels decreased the coolant velocity in the main channels, thereby reducing fluid frictional resistance losses. However, when the number of channels exceeded six, the pressure drop reduction became less pronounced. At an inlet velocity of 0.5 m·s^{-1}, the six-channel configuration showed a 7.2% pressure drop reduction compared to the four-channel design, while the eight-channel configuration only achieved a 2.4% further reduction compared to the six-channel design. Based on the comprehensive analysis of the effects of secondary channel number on both heat dissipation performance and pressure drop characteristics, the eight-channel configuration was selected for subsequent single-factor studies due to its optimal thermal characteristics and minimal inlet–outlet pressure drop.

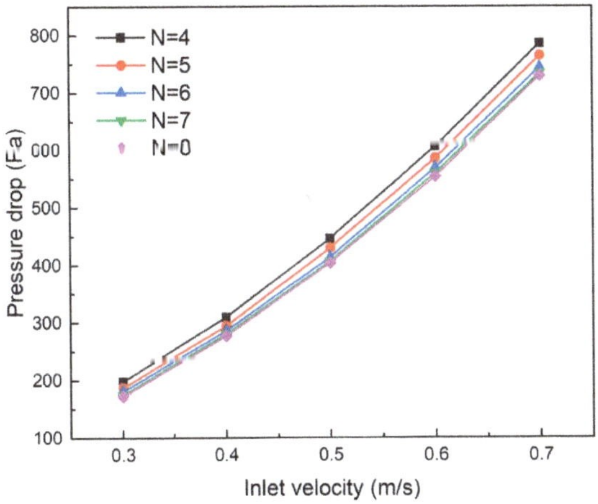

Figure 17. Effect of the number of secondary flow channels on pressure drop of cooling plates.

4.3.2. Effect of the Angle (α) of the Secondary Flow Channels on the Thermal-Hydraulic Performance of Cooling Plates

In this section, the cooling plate model was configured with eight secondary channels, with channel angles varying among 30°, 45°, 60°, 75°, and 90°, while maintaining a channel width of 1 mm and channel spacing of 2 mm. All other boundary conditions remained unchanged. The analysis focused on examining the effects of secondary channel angle (α) on both heat dissipation performance and pressure drop characteristics.

As shown in Figure 18, under the same temperature conditions, both the maximum surface temperature and temperature difference in the cooling plate exhibited an increasing trend with an increasing channel angle. At an inlet velocity of 0.5 $m \cdot s^{-1}$, the cooling plate with a 90° channel angle showed increases of 1.17 °C in maximum surface temperature and 1.18 °C in temperature difference compared to the 30° configuration. As illustrated in Figure 18c, the channel angle's impact on the average temperature was minimal. At an inlet velocity of 0.3 $m \cdot s^{-1}$, the maximum temperature difference between 90° and 30° channel angles was 0.33 °C, which decreased to merely 0.13 °C when the inlet velocity increased to 0.7 $m \cdot s^{-1}$. Furthermore, increasing the channel angle led to a higher temperature uniformity index, indicating deteriorating temperature consistency across the cooling plate surface. This phenomenon can be attributed to reduced fluid volume in secondary channels at larger angles, which decreases effective heat transfer area. Additionally, larger channel angles weaken fluid rotation and vortex intensity, reducing fluid mixing and resulting in less uniform temperature distribution within the cooling plate. However, when the channel angle exceeded 60°, the rate of change in this index decreased, indicating stabilized fluid mixing effects and thermal conduction equilibrium.

As demonstrated in Figure 19, under constant flow velocity, the inlet–outlet pressure drop decreased as the channel angle increased. This reduction occurs because larger channel angles shorten the fluid path through secondary channels while reducing vortex intensity and frictional resistance. However, this trend became less pronounced beyond 45°. At an inlet velocity of 0.5 $m \cdot s^{-1}$, the 45° channel configuration showed a 6.0% pressure drop reduction compared to the 30° configuration, while the 90° configuration only achieved a further 2.0% reduction compared to the 45° configuration. This diminishing effect occurs because fluid mixing stabilizes and pressure distribution becomes uniform at larger angles, making further angle increases less impactful on pressure drop characteristics.

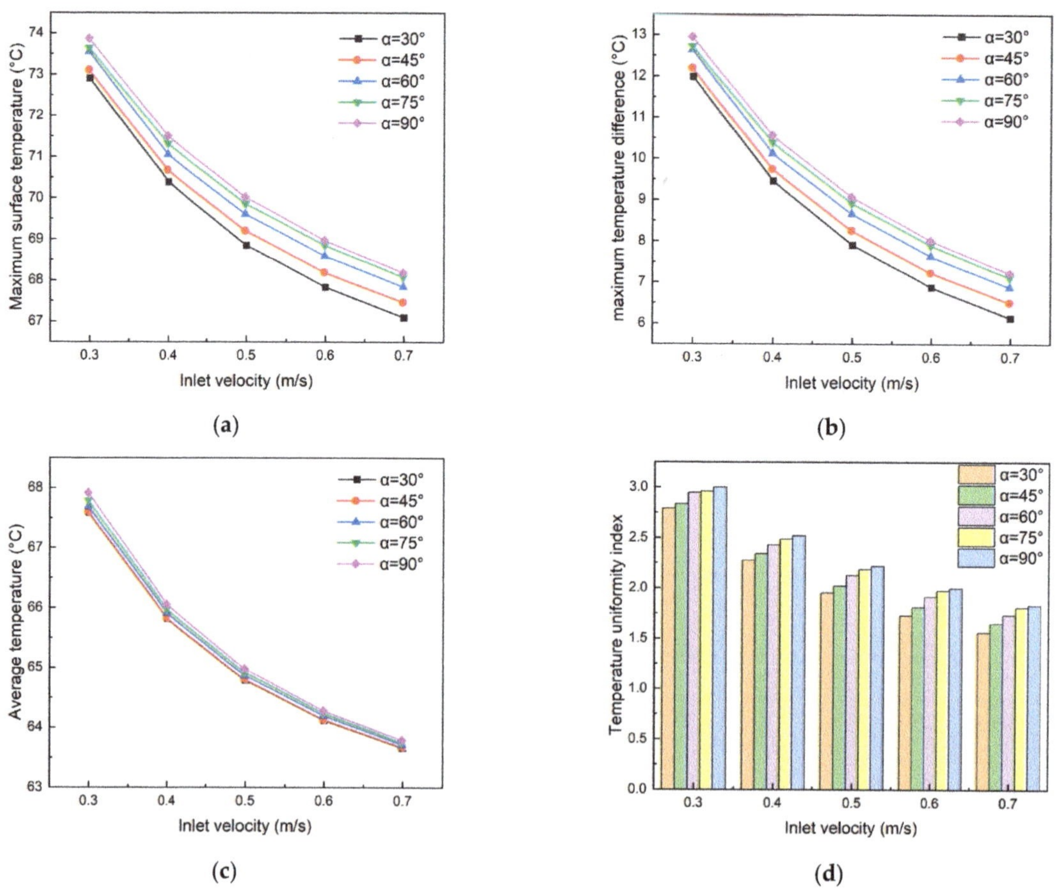

Figure 18. Effect of the angle of secondary flow channels on (**a**) maximum surface temperature, (**b**) maximum temperature difference, (**c**) average temperature, and (**d**) TUI of cooling plates.

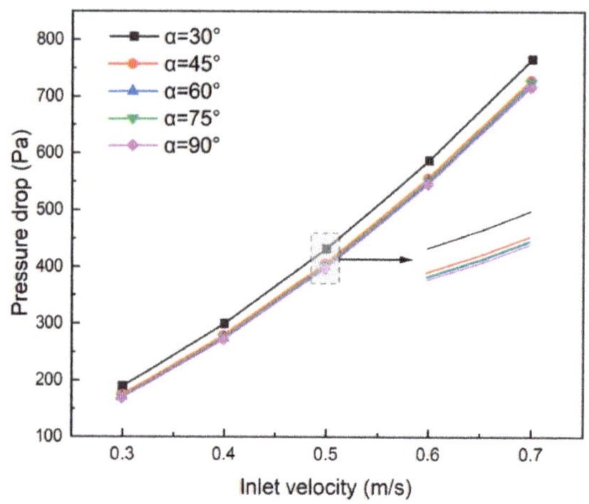

Figure 19. Effect of the angle of secondary flow channels on pressure drop of cooling plates.

Based on the comprehensive analysis of channel angle effects on both heat dissipation and pressure drop characteristics, a channel angle of 45° was selected as the optimal configuration for subsequent single-factor studies of cooling plate performance.

4.3.3. Effect of the Width (d) of Secondary Flow Channels on the Thermal-Hydraulic Performance of Cooling Plates

In this section, the cooling plate model incorporates eight secondary channels with a channel angle of 45°. The channel width varies between 0.5 mm, 0.75 mm, 1 mm, 1.25 mm, and 1.5 mm, while maintaining a constant channel spacing of 2 mm and other boundary conditions are unchanged. This analysis examines the effects of secondary channel width (*d*) on both heat dissipation performance and pressure drop characteristics.

As shown in Figure 20, under the same flow velocity, both the maximum surface temperature and temperature difference in the cooling plate increase with the channel width. At a coolant inlet velocity of 0.5 m·s^{-1}, the cooling plate with 1.5 mm channel width exhibits maximum temperature and temperature difference values 0.64 °C higher than those with 0.5 mm channel width. This phenomenon occurs because wider channels reduce fluid velocity within the channels, leading to heat stagnation or non-uniform heat transfer, thereby weakening the heat exchange capacity of the coolant. Conversely, in narrower secondary channels, dramatic local velocity variations promote vortex formation, enhancing heat exchange efficiency. As shown in Figure 20c, channel width has minimal impact on the average plate temperature, with maximum temperature variations of only 0.12 °C among different channel widths at an inlet velocity of 0.7 m·s^{-1}. Figure 20d demonstrates that at constant flow velocities (>0.4 m·s^{-1}), the temperature uniformity index increases with increasing channel width, indicating deteriorating temperature consistency. However, this effect becomes negligible beyond 1.25 mm channel width, as the fluid flow reaches a stable state and mixing effects become saturated, resulting in no further changes in heat exchange capacity or temperature uniformity.

As illustrated in Figure 21, under constant flow velocity, the inlet–outlet pressure drops decrease with increasing channel width, though at a diminishing rate. At a coolant inlet velocity of 0.5 m·s^{-1}, the pressure drop for the 1 mm channel width is 10.7% lower than that for 0.5 mm, while the 1.5 mm channel width shows only a 3.45% further reduction compared to 1 mm. This trend occurs because beyond a certain channel width, fluid velocity stabilizes within the channels, and the influence of fluid viscosity on pressure drop becomes constant, resulting in minimal pressure drop variations with further width increases.

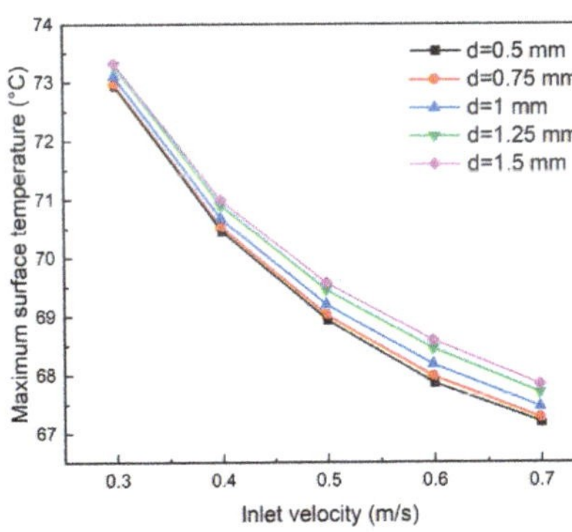

(a)

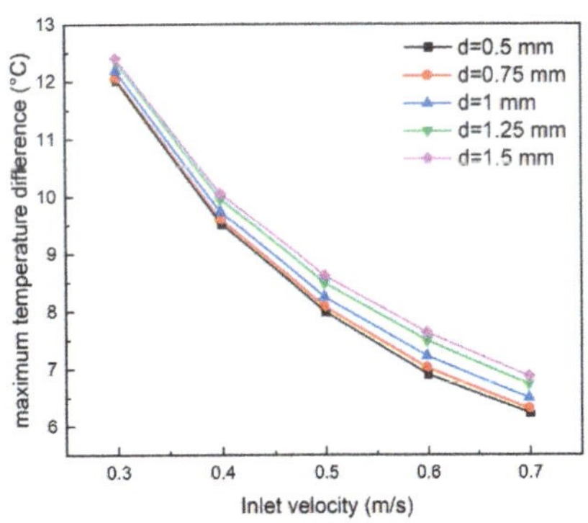

(b)

Figure 20. *Cont.*

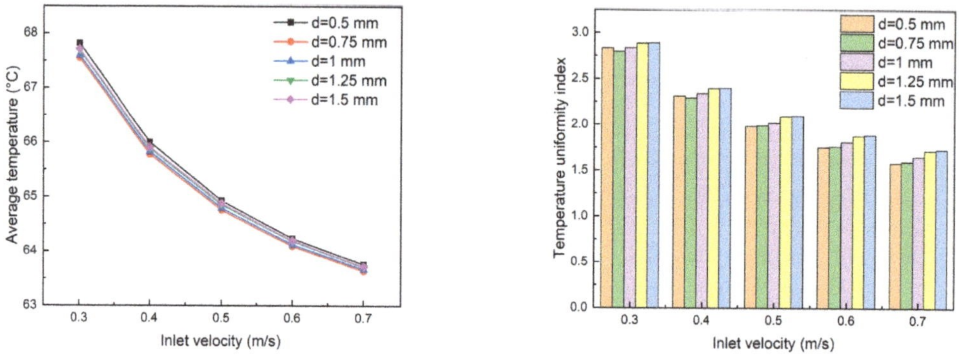

(c) (d)

Figure 20. Effect of the width of secondary flow channels on (**a**) maximum surface temperature, (**b**) maximum temperature difference, (**c**) average temperature, and (**d**) TUI of cooling plates.

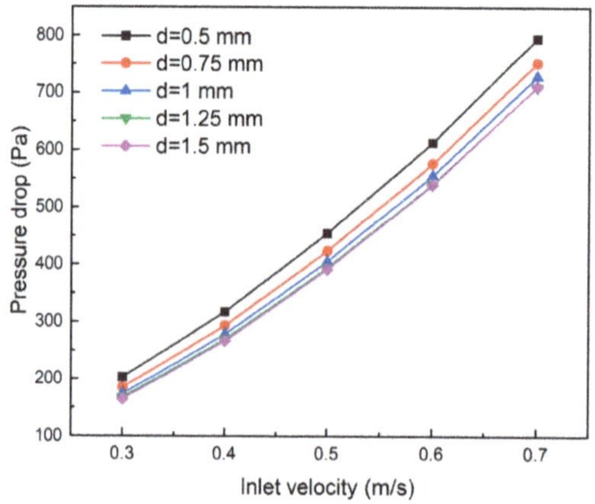

Figure 21. Effect of the width of secondary flow channels on pressure drop of cooling plates.

Based on the comprehensive analysis of secondary channel width effects on both heat dissipation performance and pressure drop characteristics, a channel width of 1 mm is selected as the optimal configuration for subsequent research, balancing optimal cooling performance with pump power efficiency.

4.3.4. Effect of the Spacing (L) of Secondary Flow Channels on the Thermal-Hydraulic Performance of Cooling Plates

In this section, the cooling plate model incorporates eight secondary channels with a channel angle of 45° and a channel width of 1 mm. The channel spacing varies between 1 mm, 1.5 mm, 2 mm, 2.5 mm, and 3 mm, while other boundary conditions remain unchanged. This analysis examines the effects of secondary channel spacing (*L*) on both heat dissipation performance and pressure drop characteristics.

As shown in Figure 22, under the same flow velocity, the heat dissipation performance improves with increasing channel spacing. At a coolant inlet velocity of 0.5 $m \cdot s^{-1}$, the cooling plate with 1 mm channel spacing exhibits maximum temperature and temperature difference values 0.65 °C higher than those with 3 mm spacing. As illustrated in Figure 22c, channel spacing has minimal impact on the average plate temperature, with the maximum temperature difference between 1 mm and 3 mm spacing configurations being only

0.22 °C at 0.5 m·s^{-1}. Figure 22d demonstrates that the temperature uniformity index decreases with increasing channel spacing, and this reduction becomes slightly more pronounced at higher inlet velocities. This improvement in temperature uniformity can be attributed to the influence of channel spacing on local flow patterns and vortex effects. At an inlet velocity of 0.3 m·s^{-1}, the difference in temperature uniformity index between 3 mm and 1 mm spacing configurations is 0.15, which increases to 0.18 at 0.7 m·s^{-1}.

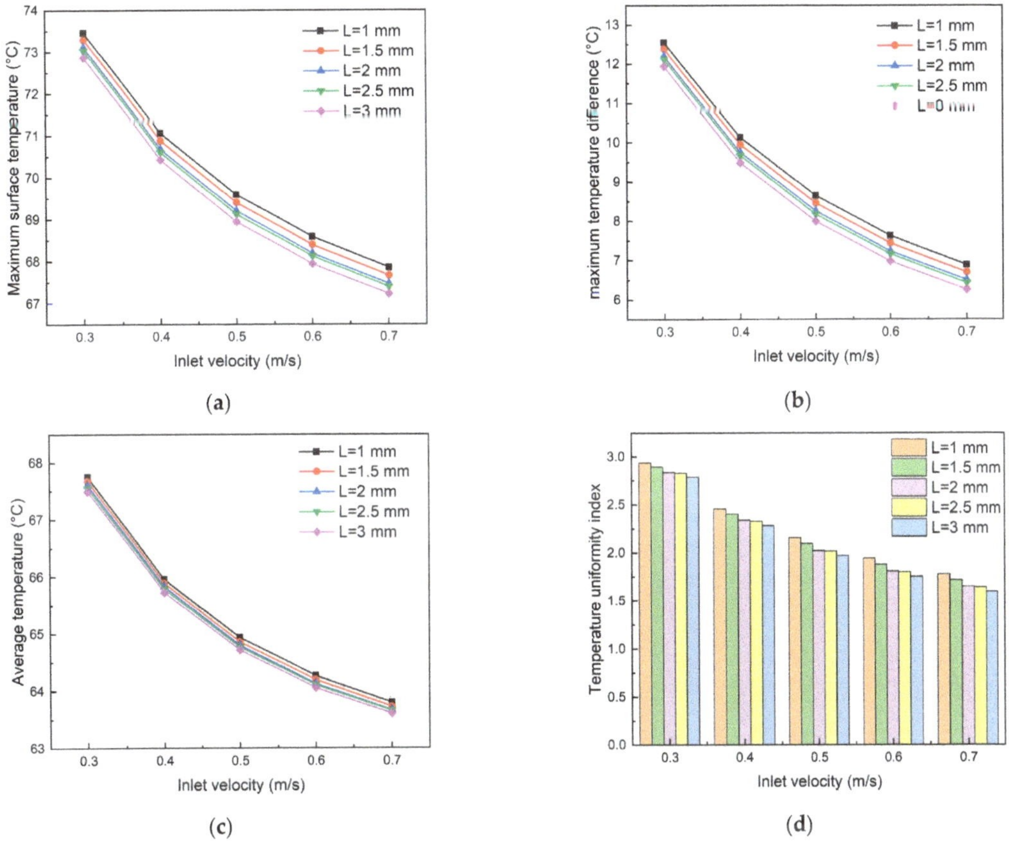

Figure 22. Effect of the spacing of secondary flow channels on (**a**) maximum surface temperature, (**b**) maximum temperature difference, (**c**) average temperature, and (**d**) TUI of cooling plates.

As shown in Figure 23, the inlet–outlet pressure drops increase with increasing coolant inlet velocity for all channel spacing configurations, with approximately consistent growth rates. From a vertical perspective, at constant inlet velocity, channel spacing variations have minimal impact on pressure drop characteristics. At an inlet velocity of 0.7 m·s^{-1}, the maximum pressure drop difference among different channel spacing configurations is only 12.39 Pa. This minimal effect occurs because changes in channel spacing do not significantly alter the flow regime within secondary channels or the frictional resistance, resulting in a minimal impact on pressure distribution.

Comprehensive analysis of secondary channel spacing effects on both heat dissipation and pressure drop characteristics reveals that while modifying channel spacing can improve heat dissipation performance to some extent, its impact on inlet–outlet pressure drop remains minimal.

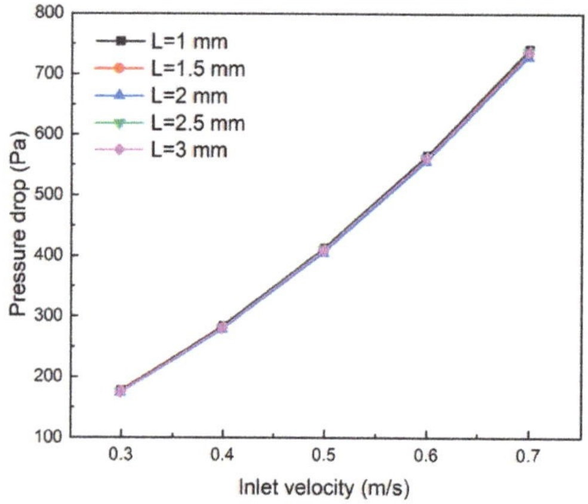

Figure 23. Effect of the spacing of secondary flow channels on pressure drop of cooling plates.

4.4. Parameter Optimization of Secondary Flow Channel Parameters Based on Orthogonal Experiment

This section employs both orthogonal experimental design and range analysis methods to examine the sensitivity and optimal parameter combinations of cooling plates under different optimization objectives when four factors interact simultaneously, thereby providing optimization strategies for PEMFC cooling plate design.

Compared to single-factor analysis, orthogonal experimental design enables a more comprehensive evaluation of multiple concurrent parameter variations, avoiding the limitations of isolated parameter studies. To investigate the effects of multiple factors on different optimization objectives, this study implements a four-factor, five-level orthogonal experimental design, comprising 25 test conditions. The factor levels are presented in Table 5, and the 25 experimental configurations are detailed in Table 6.

Table 5. Factor levels for orthogonal experiments.

Level	Factors			
	N	α	D (mm)	L (mm)
1	4	30°	0.50	1.00
2	5	45°	0.75	1.50
3	6	60°	1.00	2.00
4	7	75°	1.25	2.50
5	8	90°	1.50	3.00

Table 6. Orthogonal experimental case.

Test No.	N	α	d (mm)	L (mm)
1	4	30°	0.50	1.00
2	4	45°	0.75	1.50
3	4	60°	1.00	2.00
4	4	75°	1.25	2.50
5	4	90°	1.50	3.00
6	5	30°	0.75	2.00
7	5	45°	1.00	2.50
8	5	60°	1.25	3.00
9	5	75°	1.50	1.00
10	5	90°	0.50	1.50
11	6	30°	1.00	3.00
12	6	45°	1.25	1.00

Table 6. *Cont.*

Test No.	N	α	d (mm)	L (mm)
13	6	60°	1.50	1.50
14	6	75°	0.50	2.00
15	6	90°	0.75	2.50
16	7	30°	1.25	1.50
17	7	45°	1.50	2.00
18	7	60°	0.50	2.50
19	7	75°	0.75	3.00
20	7	90°	1.00	1.00
21	8	30°	1.50	2.50
22	8	45°	0.50	3.00
23	8	60°	0.75	1.00
24	8	75°	1.00	1.50
25	8	90°	1.25	2.00

According to the experimental design outlined above, numerical simulations were conducted using COMSOL Multiphysics for 25 cooling plate configurations with varying secondary channel structural parameters. The boundary conditions remained consistent with previous simulations, and the simulation results are presented in Table 7.

Table 7. Orthogonal simulation results at an inlet coolant velocity of 0.5 $m \cdot s^{-1}$.

Test No.	T_{max} (°)	ΔT_{max} (°)	T_{avg} (°)	U_T	P (Pa)	j_w
1	69.72	9.78	65.54	2.09	507.06	52,179.61
2	69.11	9.17	65.04	1.98	463.93	57,296.14
3	69.60	9.66	65.05	2.14	436.08	61,520.11
4	70.18	10.24	65.15	2.27	416.07	64,675.66
5	70.61	10.67	65.25	2.36	410.54	65,848.35
6	69.10	9.16	65.10	1.96	461.50	57,508.84
7	69.19	9.24	64.92	2.02	428.41	62,812.96
8	69.87	9.92	65.03	2.20	412.82	65,284.22
9	71.00	11.06	65.39	2.46	397.54	68,311.06
10	69.19	9.24	65.00	2.00	449.05	59,743.65
11	68.95	9.01	64.91	1.96	437.39	61,429.75
12	69.81	9.87	65.06	2.21	409.67	65,886.91
13	70.59	10.64	65.21	2.36	397.73	67,968.32
14	69.09	9.14	64.90	2.00	447.74	59,952.55
15	69.52	9.57	64.91	2.11	418.93	64,528.41
16	69.19	9.25	64.89	2.05	420.64	64,267.16
17	69.60	9.65	64.91	2.11	400.93	67,835.21
18	68.94	9.00	64.84	1.96	450.72	59,157.13
19	69.20	9.26	64.80	2.03	418.37	64,222.51
20	70.52	10.57	65.24	2.37	407.95	66,365.79
21	68.82	8.87	64.71	1.95	398.34	67,142.85
22	68.94	8.99	64.90	1.95	446.99	59,651.48
23	69.57	9.63	64.92	2.15	413.72	64,943.87
24	70.14	10.20	65.04	2.26	395.54	68,760.30
25	70.26	10.31	65.04	2.25	390.66	69,723.65

The range analysis method was employed to evaluate the effects of four structural parameters on both heat dissipation capacity and pressure drop characteristics of the cooling plate. This method enables the determination of primary–secondary order and optimal combinations of influencing factors on experimental indicators. The specific equations are as follows Equations (13)–(15):

$$K_{ij} = \sum_{j=1}^{4} I_{ij} \tag{13}$$

$$\overline{K_{ij}} = \frac{K_{ij}}{r} \tag{14}$$

$$\mathrm{R} = \max\left(\overline{K_{ij}}\right) - min\left(\overline{K_{ij}}\right) \tag{15}$$

where I_{ij} represents the experimental result corresponding to the factor in column i at level j, K_{ij} denotes the sum of experimental results for the factor in column i at level j, $\overline{K_{ij}}$ represents the mean value of the evaluation index sum, and R indicates the range value of factors, calculated as the difference between the maximum and minimum mean values under different levels for the factor in column i.

The range R represents the maximum difference of $\overline{K_{ij}}$ for each factor, with larger range values indicating greater influence of the corresponding factor level on the evaluation index. Therefore, the primary–secondary order of factors affecting the evaluation indices can be determined based on the magnitude of range values R.

4.4.1. Range Analysis of Maximum Surface Temperature in the Cooling Plate

Table 8 presents the range analysis results using the maximum surface temperature of the cooling plate as the evaluation index. The analysis reveals that channel width (d) has the most significant impact on the maximum surface temperature, followed by channel angle (α) and channel spacing (L), while the channel number (N) demonstrates the least influence. Through analysis of the mean value $\overline{K_{ij}}$ at different levels for each factor, where lower values indicate better performance for the maximum surface temperature index, the optimal combination of secondary channel structural parameters was determined to be $N = 7$, $\alpha = 30°$, $d = 0.5$ mm, and $L = 2.5$ mm.

Table 8. Range analysis of maximum surface temperature.

K_{ij}	N	α	d	L
K_{i1}	349.22	345.79	345.87	350.62
K_{i2}	348.34	346.64	346.50	348.22
K_{i3}	347.96	348.57	348.40	347.65
K_{i4}	347.45	349.62	349.31	346.64
K_{i5}	347.73	350.09	350.62	347.57
$\overline{K_{i1}}$	69.84	69.16	69.17	70.12
$\overline{K_{i2}}$	69.67	69.33	69.30	69.64
$\overline{K_{i3}}$	69.59	69.71	69.68	69.53
$\overline{K_{i4}}$	69.49	69.92	69.86	69.33
$\overline{K_{i5}}$	69.55	70.02	70.12	69.51
R	0.36	0.86	0.95	0.79
Factor significance	$d > \alpha > L > N$			

4.4.2. Range Analysis of Maximum Temperature Difference in the Cooling Plate

Table 9 presents the range analysis results using the maximum temperature difference across the cooling plate surface as the evaluation index. Analysis of the range values reveals that the influence pattern of various factors on the maximum temperature difference mirrors that of the maximum surface temperature. The sensitivity ranking of factors affecting the maximum temperature difference is as follows: channel width (d) > channel angle (α) >

channel spacing (L) > channel number (N). Through comparison of the mean value $\overline{K_{ij}}$ for each factor, the optimal configuration of secondary channel structural parameters for minimizing surface temperature difference was determined to be $N = 7$, $\alpha = 30°$, $d = 0.5$ mm, and $L = 2.5$ mm.

Table 9. Range analysis of maximum surface temperature difference.

K_{ij}	N	α	d	L
K_{i1}	44.52	41.07	41.16	45.91
K_{i2}	43.62	41.92	41.78	43.49
K_{i3}	43.24	43.85	43.68	42.93
K_{i4}	42.72	44.89	44.59	41.92
K_{i5}	43.01	45.37	45.89	42.85
$\overline{K_{i1}}$	8.90	8.21	8.23	9.18
$\overline{K_{i2}}$	8.72	8.38	8.36	8.70
$\overline{K_{i3}}$	8.65	8.77	8.74	8.57
$\overline{K_{i4}}$	8.54	8.98	8.92	8.38
$\overline{K_{i5}}$	8.60	9.07	9.18	8.57
R	44.52	41.07	41.16	45.91
Factor significance		$d > \alpha > L > N$		

4.4.3. Range Analysis of Temperature Uniformity Index in the Cooling Plate

Table 10 presents the range analysis results using the temperature uniformity index of the cooling plate as the evaluation index. A lower temperature uniformity index indicates more uniform temperature distribution across the cooling plate surface. Comparison of range values among the four factors reveals the sensitivity ranking of factors affecting the temperature uniformity index as follows: channel width (d) > channel angle (α) > channel spacing (L) > channel number (N). Through analysis of the mean value $\overline{K_{ij}}$ for each factor, the optimal configuration of secondary channel structural parameters for minimizing the temperature uniformity index was determined to be $N = 7$, $\alpha = 30°$, $d = 0.5$ mm, and $L = 2.5$ mm.

Table 10. Range analysis of temperature uniformity index.

K_{ij}	N	α	d	L
K_{i1}	10.84	10.01	10.01	11.28
K_{i2}	10.64	10.26	10.22	10.65
K_{i3}	10.64	10.82	10.75	10.46
K_{i4}	10.51	11.02	10.97	10.30
K_{i5}	10.56	11.09	11.24	10.51
$\overline{K_{i1}}$	2.17	2.00	2.00	2.26
$\overline{K_{i2}}$	2.13	2.05	2.04	2.13
$\overline{K_{i3}}$	2.13	2.16	2.15	2.09
$\overline{K_{i4}}$	2.10	2.20	2.19	2.06
$\overline{K_{i5}}$	2.11	2.22	2.25	2.10
R	0.06	0.22	0.25	0.20
Factor significance		$d > \alpha > L > N$		

4.4.4. Range Analysis of the Inlet–outlet Pressure Drops in the Cooling Plate

Table 11 presents the range analysis results using the inlet–outlet pressure drop of the coolant as the evaluation index. Comparison of range values among the four factors reveals the sensitivity ranking of factors affecting the pressure drop as follows: channel width (d) > channel number (N) > channel angle (α) > channel spacing (L). Through analysis of the mean value $\overline{K_{ij}}$ at different levels for each factor, the optimal configuration of secondary channel structural parameters for optimizing inlet–outlet pressure drop was determined to be $N = 8$, $\alpha = 75°$, $d = 1.5$ mm, and $L = 2.5$ mm.

Table 11. Range analysis of the inlet–outlet pressure drop.

K_{ij}	N	α	d	L
K_{i1}	2233.67	2224.91	2298.55	2135.94
K_{i2}	2149.32	2149.93	2176.45	2126.88
K_{i3}	2108.47	2111.08	2105.38	2133.92
K_{i4}	2098.62	2072.26	2049.87	2112.48
K_{i5}	2045.25	2077.13	2005.08	2126.11
$\overline{K_{i1}}$	446.73	444.98	459.71	427.18
$\overline{K_{i2}}$	429.86	429.98	435.29	425.37
$\overline{K_{i3}}$	421.69	422.21	421.07	426.78
$\overline{K_{i4}}$	419.72	414.45	409.97	422.50
$\overline{K_{i5}}$	409.05	415.43	401.02	425.22
R	37.68	29.55	58.69	4.69
Factor significance	$d > N > \alpha > L$			

4.4.5. Range Analysis of the Cooling Efficiency Coefficient in the Cooling Plate

Table 12 presents the range analysis results using the cooling efficiency coefficient of the cooling plate as the evaluation index. Comparison of range values among the four factors reveals the sensitivity ranking of factors affecting the cooling efficiency coefficient as follows: channel width (d) > channel number (N) > channel angle (α) > channel spacing (L). A higher cooling efficiency coefficient indicates enhanced heat transfer capacity per unit pumping power, and thus better cooling performance of the plate. Through analysis of the mean value $\overline{K_{ij}}$ at different factor levels, the optimal configuration of secondary channel structural parameters for maximizing the cooling efficiency coefficient was determined to be $N = 8$, $\alpha = 90°$, $d = 1.5$ mm, and $L = 2.5$ mm.

Table 12. Range analysis of the cooling efficiency coefficient.

K_{ij}	N	α	d	L
K_{i1}	301,519.88	302,528.22	290,684.42	317,687.25
K_{i2}	313,660.73	313,482.70	308,499.77	318,035.56
K_{i3}	319,765.95	318,873.64	320,888.91	316,540.36
K_{i4}	321,847.80	325,922.09	329,837.60	318,317.00
K_{i5}	330,222.14	326,209.85	337,105.79	316,436.32
$\overline{K_{i1}}$	60,303.98	60,505.64	58,136.89	63,537.45
$\overline{K_{i2}}$	62,732.15	62,696.54	61,699.96	63,607.11
$\overline{K_{i3}}$	63,953.19	63,774.73	64,177.78	63,308.07
$\overline{K_{i4}}$	64,369.56	65,184.42	65,967.52	63,663.40
$\overline{K_{i5}}$	66,044.43	65,241.97	67,421.16	63,287.26
R	5740.45	4736.33	9284.27	229.38
Factor significance	$d > N > \alpha > L$			

The parametric analysis revealed distinct relationships between structural parameters and thermal-hydraulic performance. The channel number exhibited minimal impact on thermal performance (a maximum temperature difference of only 0.1 °C between cooling plates) while significantly reducing inlet–outlet pressure drop by up to 12.7% with an increased number of channels. This suggests that optimizing the channel number can improve hydraulic performance without compromising thermal performance. Both channel angle and width demonstrated an inverse relationship between thermal and hydraulic performance: increases in either parameter degraded heat dissipation capability while reducing the pressure drop. However, a notable threshold effect was observed where pressure drop benefits diminished beyond 45° angles and 1 mm widths, indicating practical upper limits for these parameters. Channel spacing showed a positive correlation with thermal performance but negligible impact on hydraulic performance. The orthogonal experimental design provided critical insights into parameter sensitivity under simultaneous variation. For temperature-related optimization objectives, parameter sensitivity ranked as follows: $d > \alpha > L > N$. For objectives related to pressure drop and cooling efficiency coefficient, the influence ranking shifted to $d > N > \alpha > L$. These findings demonstrate that channel width (d) consistently exerts the dominant influence across all performance metrics, making it the primary design parameter for cooling plate optimization. The contrasting influence rankings between thermal and hydraulic objectives highlight the trade-offs in cooling plate design and underscore the importance of parameter-specific optimization based on application priorities in marine PEMFC systems.

5. Conclusions

In this study, six PEMFC cooling channel designs (Type A to F) were investigated through COMSOL Multiphysics simulation to analyze their thermal-hydraulic performance under varying heat flux and coolant inlet velocities. Based on the cooling efficiency coefficient evaluation, the Type B cooling plate, which demonstrated optimal comprehensive performance, was selected for secondary flow channel parameter optimization. Single-factor and orthogonal experimental designs were conducted on four parameters: number (N), angle (α), width (d), and spacing (L). The main conclusions are as follows:

(1) Under identical heat flux conditions, the Type C cooling plate exhibited the best thermal performance, while Type A's performance was the poorest. When heat flux increased from 0.107 $W \cdot cm^{-2}$ to 0.404 $W \cdot cm^{-2}$, the Type A cooling plate showed the largest increases in maximum surface temperature (10.59 °C) and temperature difference (10.65 °C). However, after incorporating secondary flow channels, the Type B cooling plate demonstrated smaller increases in maximum surface temperature (9.02 °C) and temperature difference (9.07 °C). Types C and E cooling plates exhibited excellent temperature uniformity, with minimal changes in their temperature uniformity index as heat flux increased, while Type A showed the largest increase (2.26). Compared to Type A, the increase in Type B was reduced to 2.10, indicating that the incorporation of secondary flow channels improved the thermal performance of the parallel channel (Type A).

(2) The incorporation of secondary flow channels improved thermal performance only for parallel channels but significantly reduced pressure drops in conventional channels, thereby enhancing their cooling efficiency coefficient. Although the Type C cooling plate showed the best thermal performance, its pressure drop reached 25,542 Pa at 0.7 $m \cdot s^{-1}$, one to two orders of magnitude higher than the other five channels, resulting in the lowest cooling efficiency coefficient (0.08×10^4). In contrast, the Type B cooling plate achieved the highest cooling efficiency coefficient (2.66×10^4) due to its low pressure drop (721 Pa). The pressure drop of Type B was 28.2% lower than Type A (171 vs. 238 Pa at 0.3 $m \cdot s^{-1}$).

(3) The structural parameters of secondary flow channels exhibited varying degrees of influence on cooling plate performance. The number of channels (N) had minimal impact on thermal performance (with a maximum temperature difference of only 0.1 °C) but significantly reduced inlet–outlet pressure drops (up to 12.7%). Increasing channel angle (α) deteriorated thermal performance, while reducing the pressure drop showed diminishing improvements beyond 45°. Increased channel width (d) similarly

reduced thermal performance and pressure drop, though with diminishing returns (pressure drop decreased by 10.7% from 0.5 mm to 1 mm, but only 3.45% from 1 mm to 1.5 mm at 0.5 $m{\cdot}s^{-1}$). Increased channel spacing (L) slightly improved thermal performance but had minimal impact on pressure drop.

(4) For the cooling plate's optimization objectives of thermal performance (maximum surface temperature, maximum temperature difference, and temperature uniformity index), the factor influence ranking is $d > \alpha > L > N$ (with the optimal combination being $N = 7$, $\alpha = 30°$, $d = 0.5$ mm, and $L = 2.5$ mm); for pressure drop optimization, the ranking is $d > N > \alpha > L$ (with the optimal combination being $N = 8$, $\alpha = 75°$, $d = 1.5$ mm, and $L = 2.5$ mm); and for cooling efficiency coefficient optimization, the ranking is $d > N > \alpha > L$ (with optimal combination: $N = 8$, $\alpha = 90°$, $d = 1.5$ mm, and $L = 2.5$ mm).

This study focused solely on the thermal dissipation performance of PEMFC cooling plates without addressing their low-temperature heating functionality in PEMFC thermal management systems. This aspect could be considered in future research under low-temperature heating conditions, ultimately contributing to more efficient thermal-hydraulic designs for marine PEMFC applications.

Author Contributions: W.C., Conceptualization, Supervision, Project administration, Methodology, Funding acquisition, Writing—Review and Editing; Y.Y., Visualization, Validation, Investigation, Formal analysis, Data curation, Writing—Original Draft; Y.L., Validation, Investigation, Writing—Original Draft. All authors have read and agreed to the published version of the manuscript.

Funding: This research was funded by Shanghai Industrial Collaborative Innovation (Technology) Project, China (grant number: XTCX-KJ-2023-29).

Data Availability Statement: The raw data supporting the conclusions of this article will be made available by the authors on request.

Conflicts of Interest: This work represents original research that has not been previously published in any form, nor is it currently under consideration for publication elsewhere. No conflict of interest exists in the submission of this manuscript, and the manuscript was approved for publication by all authors.

Abbreviations

The following abbreviations are used in this manuscript:

PEMFC	Proton exchange membrane fuel cell.
PEMFCs	Proton exchange membrane fuel cells.
CFD	Computational Fluid Dynamics.
IMO	International Maritime Organization.
SMCHS	Serpentine mini-channel heat sinks.
HOR	Hydrogen oxidation reaction.
ORR	Oxygen reduction reaction.
MEA	Membrane electrode assemblies.
TUI	Temperature uniformity index.

References

1. Nivolianiti, E.; Karnavas, Y.L.; Charpentier, J.F. Energy management of shipboard microgrids integrating energy storage systems: A review. *Renew. Sustain. Energy Rev.* **2024**, *189*, 114012. [CrossRef]
2. Bilgili, L.; Ölçer, A.I. IMO 2023 strategy-Where are we and what's next? *Mar. Policy* **2024**, *160*, 105953. [CrossRef]
3. Hamedani, E.A.; Alenabi, S.A.; Talebi, S. Hydrogen as an energy source: A review of production technologies and challenges of fuel cell vehicles. *Energy Rep.* **2024**, *12*, 3778–3794. [CrossRef]
4. Sürer, M.G.; Arat, H.T. Advancements and current technologies on hydrogen fuel cell applications for marine vehicles. *Int. J. Hydrogen Energy* **2022**, *47*, 19865–19875. [CrossRef]

5. Inal, O.B.; Deniz, C. Assessment of fuel cell types for ships: Based on multi-criteria decision analysis. *J. Clean. Prod.* **2020**, *265*, 121734. [CrossRef]
6. Chen, Z.X.; Ingham, D.B.; Ismail, M.S.; Ma, L.; Hughes, K.J.; Pourkashanian, M. Dynamics of liquid water in the anode flow channels of PEM fuel cells: A numerical parametric study. *J. Energy Inst.* **2019**, *92*, 1956–1967. [CrossRef]
7. Zhang, Z.; Mao, J.; Liu, Z. Advancements and insights in thermal and water management of proton exchange membrane fuel cells: Challenges and prospects. *Int. Commun. Heat Mass Transf.* **2024**, *153*, 107376. [CrossRef]
8. Ramezanizadeh, M.; Nazari, M.A.; Ahmadi, M.H.; Chen, L. A review on the approaches applied for cooling fuel cells. *Int. J. Heat Mass Transf.* **2019**, *139*, 517–525. [CrossRef]
9. Zhang, G.; Kandlikar, S.G. A critical review of cooling techniques in proton exchange membrane fuel cell stacks. *Int. J. Hydrogen Energy* **2012**, *37*, 2412–2429. [CrossRef]
10. Bargal, M.H.S.; Abdelkareem, M.A.A.; Tao, Q.; Li, J.; Shi, J.; Wang, Y. Liquid cooling techniques in proton exchange membrane fuel cell stacks: A detailed survey. *Alex. Eng. J.* **2020**, *59*, 635–655. [CrossRef]
11. Amirfazli, A.; Asghari, S.; Sarraf, M. An investigation into the effect of manifold geometry on uniformity of temperature distribution in a PEMFC stack. *Energy* **2018**, *145*, 141–151. [CrossRef]
12. Liu, F.; Chen, Y.; Qin, W.; Li, J. Optimal design of liquid cooling structure with bionic leaf vein branch channel for power battery. *Appl. Therm. Eng.* **2023**, *218*, 119283. [CrossRef]
13. Tong, G.; Xu, X.; Yuan, Q.; Yang, Y.; Tang, W.; Sun, X. Research on influencing factors of heat transfer enhancement fins in fuel cell cooling channel. *Ionics* **2021**, *27*, 743–757. [CrossRef]
14. Chen, X.; Chai, F.; Hu, S.; Tan, J.; Luo, L.; Xie, H.; Wan, Z.; Qu, Z. Design of PEMFC bipolar plate cooling flow field based on fractal theory. *Energy Convers. Manage. X* **2023**, *20*, 100445. [CrossRef]
15. Guo, R.; Li, L. Heat dissipation analysis and optimization of lithium-ion batteries with a novel parallel-spiral serpentine channel liquid cooling plate. *Int. J. Heat Mass Transf.* **2022**, *189*, 122706. [CrossRef]
16. Ebrahimi, F.; Kermani, M.J. Generalization of the method of flow channel blocking in PEM fuel cells; extensions from straight-parallel to parallel-serpentine flow fields. *Int. J. Hydrogen Energy* **2025**, *in press*. [CrossRef]
17. Chen, F.C.; Gao, Z.; Loutfy, R.O.; Hecht, M. Analysis of optimal heat transfer in a PEM fuel cell cooling plate. *Fuel Cells* **2003**, *3*, 181–188. [CrossRef]
18. Hu, G.; Fan, J.; Chen, S.; Liu, Y.; Cen, K. Three-dimensional numerical analysis of proton exchange membrane fuel cells (PEMFCs) with conventional and interdigitated flow fields. *J. Power Sources* **2004**, *136*, 1–9. [CrossRef]
19. Wang, X.D.; Duan, Y.Y.; Yan, W.M.; Peng, X.F. Effects of flow channel geometry on cell performance for PEM fuel cells with parallel and interdigitated flow fields. *Electrochim. Acta* **2008**, *53*, 5334–5343. [CrossRef]
20. Acar, M.C. Modeling the influence of coolant flow directions on thermal performance of PEM fuel cell cooling plates with serpentine and straight flow channels. *Thermochim. Acta* **2023**, *726*, 179560. [CrossRef]
21. Yu, S.H.; Sohn, S.; Nam, J.H.; Kim, C.J. Numerical study to examine the performance of multi-pass serpentine flow-fields for cooling plates in polymer electrolyte membrane fuel cells. *J. Power Sources* **2009**, *194*, 697–703. [CrossRef]
22. Baek, S.M.; Yu, S.H.; Nam, J.H.; Kim, C.J. A numerical study on uniform cooling of large-scale PEMFCs with different coolant flow field designs. *Appl. Therm. Eng.* **2011**, *31*, 1427–1434. [CrossRef]
23. Imran, A.A.; Mahmoud, N.S.; Jaffal, H.M. Numerical and experimental investigation of heat transfer in liquid cooling serpentine mini-channel heat sink with different new configuration models. *Therm. Sci. Eng. Prog.* **2018**, *6*, 128–139. [CrossRef]
24. Wang, Y.; Sun, Z.Y.; Yang, L. Optimizing temperature distribution in a PEMFC stack: A computational study on cooling plate and coolant dynamics. *Int. J. Hydrogen Energy* **2025**, *97*, 88–103. [CrossRef]
25. Deng, T.; Zhang, G.; Ran, Y. Study on thermal management of rectangular Li-ion battery with serpentine-channel cold plate. *Int. J. Heat Mass Transf.* **2018**, *125*, 143–152. [CrossRef]
26. Hamrang, A.; Abdollahzadeh, M.; Bilondi, A.M.; Bagherighajari, F.; Rahgoshay, S.M.; Pascoa, J.C. Comparison of PEMFC performance with parallel serpentine and parallel serpentine-baffled flow fields under various operating and geometrical conditions; a parametric study. *Int. J. Hydrogen Energy* **2023**, *48*, 7442–7459. [CrossRef]
27. Jaffal, H.M.; Mahmoudn, N.S.; Imran, A.A.; Hasan, A. Performance enhancement of a novel serpentine channel cooled plate used for cooling of Li-ion battery module. *Int. J. Therm. Sci* **2023**, *184*, 107955. [CrossRef]
28. González-Morán, L.; Suárez, C.; Iranzo, A.; Han, L.; Rosa, F. A numerical study on heat transfer for serpentine-type cooling channels in a PEM fuel cell stack. *Energy* **2024**, *307*, 132634. [CrossRef]
29. Sheng, L.; Su, L.; Zhang, H.; Li, K.; Fang, Y.; Ye, W.; Fang, Y. Numerical investigation on a lithium-ion battery thermal management utilizing a serpentine-channel liquid cooling plate exchanger. *Int. J. Heat Mass Transf.* **2019**, *141*, 658–668. [CrossRef]
30. Atyabi, S.A.; Afshari, E.; Zohravi, E.; Udemu, C.M. Three-dimensional simulation of different flow fields of proton exchange membrane fuel cell using a multi-phase coupled model with cooling channel. *Energy* **2021**, *234*, 121247. [CrossRef]
31. Sasmito, A.P.; Kurnia, J.C.; Mujumdar, A.S. Numerical evaluation of various gas and coolant channel designs for high performance liquid-cooled proton exchange membrane fuel cell stacks. *Energy* **2012**, *44*, 278–291. [CrossRef]

32. Fan, L.; Li, J.; Chen, Y.; Zhou, D.; Jiang, Z.; Sun, J. Study on the cooling performance of a new secondary flow serpentine liquid cooling plate used for lithium battery thermal management. *Int. J. Heat Mass Transf.* **2024**, *218*, 124711. [CrossRef]
33. Arshad, A.; Ali, H.M.; Habib, A.; Bashir, M.A.; Jabbal, M.; Yan, Y. Energy and exergy analysis of fuel cells: A review. *Therm. Sci. Eng. Prog.* **2019**, *9*, 308–321. [CrossRef]
34. Chen, X.; Liu, Q.; Fang, Y.; He, L.; Huang, T.; Zhang, Y.; Wan, Z.; Wang, X. Numerical study of a MIMO-shaped cooling plate in PEMFC stack for heat transfer enhancement. *Energy Rep.* **2021**, *7*, 5804–5814. [CrossRef]
35. Larminie, J.; Dicks, A.; McDonald, M.S. *Fuel Cell Systems Explained*; John Wiley: Chichester, UK, 2003.
36. Tiwaey, B.; Kumar, R.; Singh, P.K. Thermofluidic characteristic of a nanofluid-cooled oblique fin heat sink: An experimental and numerical investigation. *Int. J. Therm. Sci.* **2022**, *171*, 107214.
37. Afshari, E.; Ziaei-Rad, M.; Dehkordi, M.M. Numerical investigation on a novel zigzag-shaped flow channel design for cooling plates of PEM fuel cells. *J. Energy Inst.* **2017**, *90*, 752–763. [CrossRef]
38. Zhang, X.; Wang, J.; Cai, J.; He, Z.; Tian, H.; Shu, G.; Shi, L. Experimental study on operating parameters matching characteristic of the organic Rankine cycle for engine waster heat recovery. *Energy* **2021**, *244*, 122681. [CrossRef]

Journal of
Marine Science
and Engineering

MDPI

Article

A Numerical Study for the Self-Propulsion Performance of a Propulsion System Using the Coanda Effect

Jun-Hee Lee [1], Kwang-Jun Paik [1,*] and Do-Han Oh [2]

[1] Department of Naval Architecture and Ocean Engineering, Inha University, Incheon 22201, Republic of Korea; harry049@naver.com
[2] Department of Naval Architecture and Ocean Engineering, Hongik University, Sejong 30016, Republic of Korea; dohanoh@hongik.ac.kr
* Correspondence: kwangjun.paik@inha.ac.kr; Tel.: +82-32-860-7331

Abstract: This study evaluated the performance of a ship propeller numerically using the Coanda effect. The simulations applied a model based on a 6.5K DWT tanker and conducted self-propulsion assessments for three types of propellers: the original propeller, a normal propeller, and a Coanda propeller. The numerical simulations used the unsteady Reynolds-averaged Navier–Stokes (URANS) equations, incorporating the SST k–ω turbulence model. The influence of the additional thrust generated by the Coanda effect on the hull resistance and self-propulsion factors was analyzed. The key findings showed that the Coanda-based propeller achieved efficient propulsion performance by generating additional lift even at low rotational speeds. A self-propulsion analysis showed that the Coanda propeller required approximately 7.8% less delivered power than the original propeller. These results suggest that propulsion systems utilizing the Coanda effect offer superior efficiency and economic advantages over traditional technologies. This study provides critical baseline data for assessing the feasibility of a Coanda propeller, with further validation planned through full-scale ship simulations.

Keywords: Coanda effect; high-efficiency propeller; self-propulsion; unsteady Reynolds averaged Navier–Stokes (URANS); computational fluid dynamics (CFD)

1. Introduction

With the recent tightening of environmental regulations by the International Maritime Organization (IMO) and the European Union (EU), there has been an urgent need for technological solutions to improve energy efficiency and reduce carbon emissions in shipping. Regulations such as the energy efficiency existing ship index (EEXI) and carbon intensity indicator (CII) introduced by the IMO are driving the development of various technologies to reduce fuel consumption and improve the efficiency of existing ships. The EU aims to accelerate the decarbonization of the maritime sector by implementing the 'Fuel EU Maritime' regulation starting in 2025. These regulations necessitate technological innovation to reduce fuel consumption and enhance energy efficiency, promoting new technologies across ship design and operations [1]. The Fuel EU Maritime regulation provides a foundational assessment of greenhouse gas reduction strategies and the economic feasibility of fuel transition in the shipping industry [2].

Traditional ship propulsion systems have undergone extensive development and optimization over a long period. Considerable research has focused on integrating additional devices, such as energy-saving devices (ESDs) [3–5], propeller boss cap fins (PBCFs) [6–8], and rudder bulbs, to reduce fuel consumption and enhance propulsion performance.

Su et al. [3] investigated an ESD consisting of a pre-swirl stator and a rudder bulb through experimental and numerical analyses on a cargo ship. They reported that the ESD positively influenced the propulsion performance by enhancing the interactions between the hull, rudder, propeller, and ESD. Improvements in the wake flow and homogenization of hub vortices were observed, highlighting the ability to improve the propulsion performance while maintaining stable operating conditions. Gaggero [6] used computational fluid dynamics (CFD) to improve the design of the propeller boss cap fin (PBCF) device. They showed that the improvements in efficiency were due primarily to the reduced hub vortex strength, minimization of net torque, and the effect of additional fins on the blade performance.

The limitations of existing technologies, including the limited applicability, design and installation complexity, and physical efficiency constraints, require adopting new design concepts and innovative technologies. This paper proposes a ship propulsion system utilizing the Coanda effect to solve these challenges. The Coanda effect, a phenomenon that leverages the behavior of a fluid flowing along a curved surface to control flow, has been proven effective across various industries, including the aerospace industry.

Xu et al. [9] investigated the application of the steady Coanda effect to reduce the aerodynamic drag of a ship using Large Eddy Simulation (LES). This study demonstrated that Coanda jet injection could effectively control the wake flow, resulting in drag reduction and improved energy efficiency.

Pfingsten and Radespiel [10] investigated jet blowing at the flap to delay the onset of wing stall and enhance lift performance. They also examined the changes in pressure distribution caused by variations in the jet injection rates and angles of attack, providing insights into the influence of jet flow on the aerodynamic behavior of the wing. Burnazzi et al. [11] conducted numerical simulations of a high-lift airfoil equipped with a Coanda flap to compare the lift and pressure based on the design parameters. They explored the effects of variations in the camber, thickness, droop nose, and jet momentum coefficients to identify the parameters effective for significant lift generation and stall delays in aerodynamic performance. Lee et al. [12] reported a novel type of unmanned aerial vehicle (UAV) using the Coanda effect for lift. Optimized surface geometries were designed by analyzing the flow around the body through experiments and simulations to maximize lift. The Coanda effect significantly reduced energy consumption compared to a conventional multicopter.

Jun et al. [13] examined how the Coanda effect affects the turning performance of a ship by applying it to the rudder. The effects of the rudder angle and jet momentum coefficient on the turning performance of a rudder using the Coanda effect were analyzed using model tests. The results revealed a tendency for advanced distance and turning diameter to decrease as the rudder angle and jet momentum coefficient increased. Furthermore, the turning performance of the rudder was most effective at specific coefficient values. Maltsev et al. [14] designed a hydrofoil capable of jet injection and conducted model experiments in a cavitation tunnel, confirming that the lift increased as the jet injection volume increased, achieving an up to threefold increase in lift. Their findings suggested the potential application of this technology to rudders. Furthermore, jet injection experiments with a rotor attached to the leading edge of the hydrofoil [15] and numerical analyses [16] showed that the rotation of the rotor enhanced the circulation flow of the jet, increasing the lift generated by the hydrofoil. In addition, previous studies confirmed that jet injections control cavitation and enhance hydrofoil lift [17–20].

Furthermore, a ship propulsion system using the Coanda effect was developed by initially implementing the concept on a two-dimensional hydrofoil that was subsequently extended to a three-dimensional propeller configuration for performance evaluation.

Eom et al. [21,22] aimed to improve the performance of hydrofoils by implementing jet injection based on the NACA0012. They confirmed that the lift increased at the same angle of attack because of the Coanda effect induced by jet injection. In addition, using a newly developed cross-sectional geometry, the effects of the slit height, tip height, and slit position on the performance of jet-injected hydrofoils were analyzed using model experiments and simulations. Lee et al. [23] extended the present study by designing a three-dimensional jet-injected propeller. They validated a numerical scheme using model tests and simulations and reported that additional thrust was generated due to the Coanda effect. In addition, torque was reduced, and propeller efficiency increased by approximately 8.7%. Lee et al. [24] focused on optimizing the slit geometry for jet injection using numerical simulations. The parameters defining the slit geometry (hydrofoil geometry, cover length, slit position, and slit height) and jet injection patterns (slit-tunnel area ratio and slit type) were defined. Performance analyses of thrust, torque, and efficiency across these parameters showed that the optimized slit geometry achieved approximately 10.8% and 2% efficiency improvement compared to the conventional propeller and initial model, respectively. In addition, a significant reduction in pressure drop regions was achieved compared to the initial design. Previous studies evaluated the performance under uniform flow conditions, which failed to capture the non-uniform wake (i.e., effective wake) flow characteristics generated behind the stern. Since ship propellers operate within these intricate wake environments, performance evaluations must consider such wake effects to ensure realistic propulsion efficiency assessments.

The performance evaluations must consider such wake effects to ensure realistic propulsion efficiency assessments. This study assessed the feasibility of a new ship propulsion system by applying a Coanda effect-based propeller to ship propulsion. The required power, propulsion efficiency, and potential of the Coanda propeller were evaluated and compared with those of the original propeller, providing insights into its practical implementation in future ship propulsion systems.

2. Numerical Set-Up

2.1. Geometries of Target Ship and Propellers

A 6.5K DWT tanker ship was used as the target ship, as shown in Figure 1. A scale ratio of 20 was used. Table 1 lists the main specifications. The length between perpendiculars (L_{PP}) was 102 m, and the draft in design condition was 6.65 m. The target speed was 13.5 knots. The ship included a rudder bulb and flow control fin as energy-saving devices.

Figure 1. Geometry of the 6.5K DWT tanker.

The target propeller was a KP1364 propeller with skew and rake, as listed in Table 2. The practicality of marine propellers utilizing the Coanda effect was investigated by applying a nozzle shape based on the geometric characteristics of the original propeller. Figure 2 shows the geometry of the propeller designed using the initial design. The initial design used a NACA 66-412 blade cross-section (Figure 3), maintaining the pitch and chord ratios of the original propeller design but developed without incorporating skew and rake. The KP1364 propeller of the target ship is referred to as the original propeller, which was used in the validation and verification study for the open-water test. The propeller without a nozzle slit is defined as the normal propeller, while the propeller with an applied nozzle slit is defined as the Coanda propeller.

Table 1. Main specifications of the objective ships (6.5K DWT tanker).

Main Specifications	Full Scale	CFD Model
Ship speed (Fr)	0.219	
Scale ratio	1	20
Length between perpendiculars, L_{PP} (m)	102	5.10
Breadth, B_{WL} (m)	18.6	0.93
Draft, d (m)	6.65	0.3325
Displacement, ∇ (m^3)	9366	1.171
Vertical center of gravity, KG (m)	2672	6.680
Moment of inertia, I_{xx}/B, I_{yy}/L_{PP}, I_{zz}/L_{PP}	0.40, 0.25, 0.25	

Table 2. Propeller principal particulars.

Main Specifications	KP1364
Full-scale diameter (m)	3.600
Model-scale diameter (m)	0.180
Pitch ratio, P/D_P ($0.7R_P$)	0.632
Thickness ratio, t/D ($0.7R_P$)	0.044
Chord ratio, c/D_P ($0.7R_P$)	0.300
Number of blades	4
Hub ratio	0.180
Direction of rotation	Right-handed

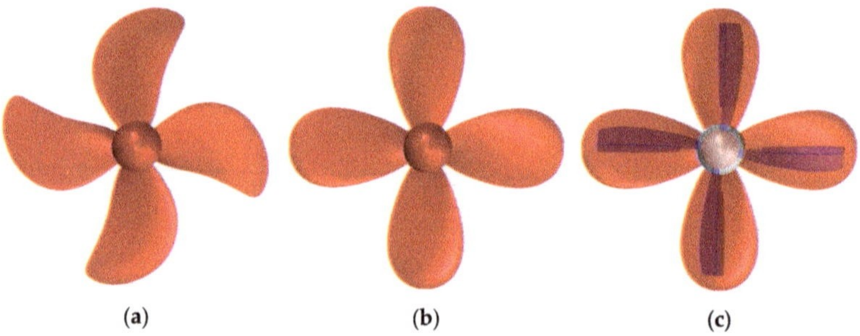

Figure 2. Geometry of three propellers: (**a**) original propeller; (**b**) normal propeller; (**c**) Coanda propeller.

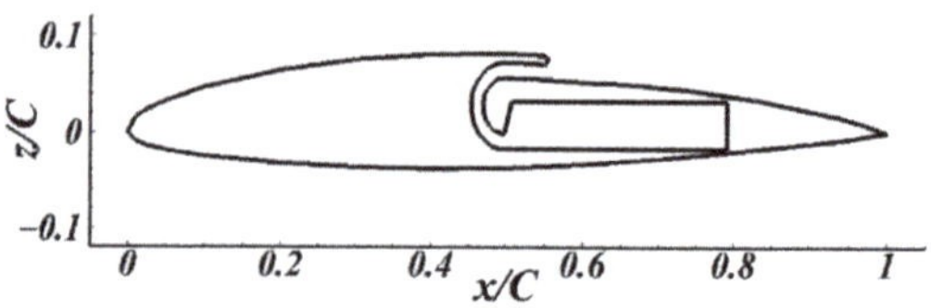

Figure 3. Geometry of the cross-section of the applied nozzle slit (adapted from [24]).

2.2. Governing Equations

The simulations were carried out using version 16.06 of STAR-CCM+. The governing equations were based on the unsteady Reynolds averaged Navier–Stokes (URANS) equations to describe the flow field.

$$\frac{\partial \bar{u}_i}{\partial \bar{x}_i} = 0 \quad (1)$$

$$\frac{\partial \bar{u}_i}{\partial t} + \bar{u}_j \frac{\partial \bar{u}_i}{\partial \bar{x}_j} = -\frac{1}{\rho}\frac{\bar{p}}{\bar{x}_j} + \frac{1}{\rho}\frac{\partial^2}{\partial \bar{x}_j}\left(\mu \frac{\partial \bar{u}_i}{\partial \bar{x}_j} - \rho\overline{u_i' u_j'}\right) \quad (2)$$

where ρ, v, p, x_i (x, y, z), and u_i (u, v, w) represent the fluid density, kinematic coefficient, pressure, and the coordinates and the fluid velocity components in the direction of each coordinate axis, respectively.

2.3. Numerical Scheme

The volume of fluid (VOF) method was adopted to track the interface between the air and water phases. The shear stress transport (SST) k–ω method [25] was used for the turbulence model. All y^+ treatments were used for the boundary layer. The SIMPLE algorithm was used for pressure–velocity coupling and second-order time implicit scheme to reach convergence for a given physical time. In the analysis, the under-relaxation factors of velocity and pressure of 0.6 and 0.4, respectively, were used.

The free motion of dynamic fluid–body interaction (DFBI) model was used to implement the ship motion. The simulation considered the 2DOF (free heave and pitch), and the overset mesh was used owing to ship motion. The sliding mesh was used in all simulations to model the actual propeller.

2.4. Computation Domain and Grid System

Figure 4 presents the computational domains for resistance, propeller open water (POW), and self-propulsion simulations. Computational domain size is recommended to minimize boundary effects in simulations [26,27].

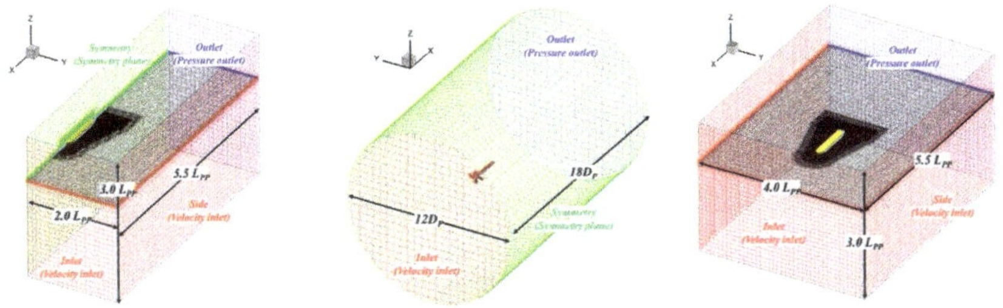

(a) (b) (c)

Figure 4. Geometry of three propellers: (**a**) resistance domain; (**b**) POW domain; (**c**) self-propulsion domain (yellow part: hull).

The boundary conditions were set to the velocity inlet and pressure outlet. The wave damping method provided by STAR-CCM+ [28] was used to prevent wave reflections at the computational domain boundaries in the CFD simulation.

The half-width computational domain with symmetry boundary conditions was used for the resistance simulation, and the number of grids was 2.97 M. Because of the need to simulate the propeller directly, the self-propulsion simulation was used for the full body, and the number of grids was 12.3 M. In the POW simulation, 3.78 M grids were used. Figure 5 shows the grid distribution around the hull and propeller. The coordinate system used in the calculation, where x, y, and z are the bow direction, port side, and deck direction, was the positive direction. Three types of propellers were attached to the stern, with jet tubes positioned along the shaft, as shown in Figure 6.

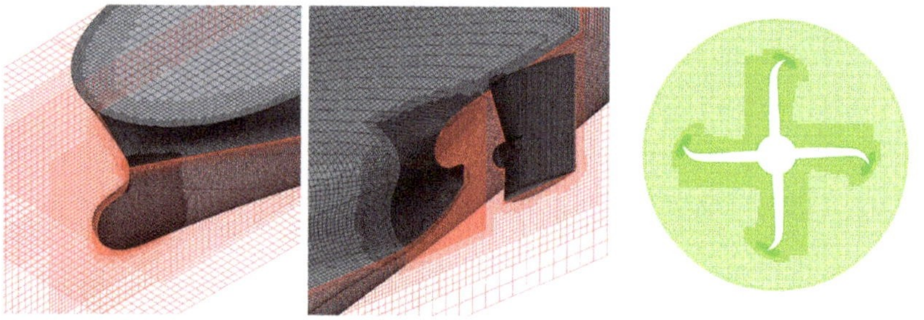

(a) (b) (c)

Figure 5. Grid distribution for the resistance and POW simulation: (**a**) perspective view (bow); (**b**) perspective view (bow); (**c**) stern view (propeller).

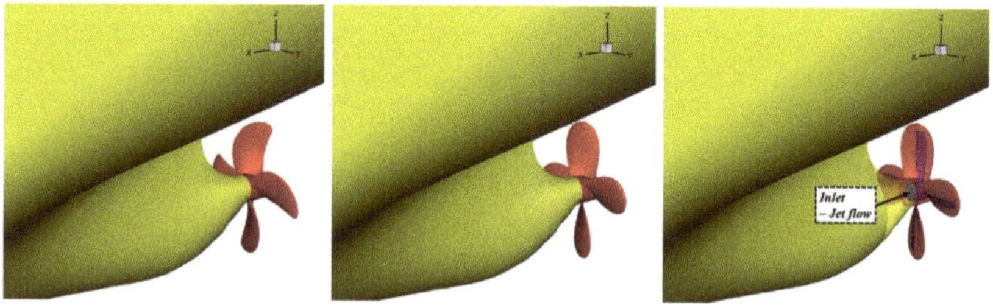

(a) (b) (c)

Figure 6. Geometry of the stern and propellers for a 6.5K DWT tanker: (**a**) original propeller; (**b**) normal propeller; (**c**) Coanda propeller.

2.5. Verification and Validation

2.5.1. Verification

A convergence test was performed to verify the results. The resistance and POW simulations were calculated independently. The evaluation was performed using the grid convergence index (GCI), which quantifies the degree of convergence in numerical analyses [29]. A convergence test based on the grid size was conducted, followed by an analysis of the time step conditions for the resistance simulation. The grid conditions for resistance were categorized as fine, medium, and coarse, corresponding to cell counts of 5,697,248, 2,976,710, and 1,553,994, respectively (Table 3). The grid condition for the POW test was set to cell counts of 7,068,215, 3,787,544, and 2,028,383.

Table 3. Grid convergence test conditions.

Variables	Fine Grid	Medium Grid	Coarse Grid
Number of grids (at resistance test)	5,697,248	2,976,710	1,553,994
Number of grids (at POW test)	7,068,215	3,787,544	2,028,383
Time step [s] (at resistance test)	0.005	0.010	0.025

The ratio of the differences in physical quantities, ε_{21} and ε_{32}, was used to define the convergence ratio, R_i. r is the grid refinement ratio. The following evaluation criteria were applied based on the convergence ratio [29,30].

$$R_i = \varepsilon_{21}/\varepsilon_{32} \tag{3}$$

$$\begin{aligned} &(i)\ \textit{Monotonic convergence}:\ 0 < R_i < 1 \\ &(ii)\ \textit{Oscillatory convergence}:\ R_i < 0 \\ &(iii)\ \textit{Divergence}:\ R_i > 1 \end{aligned} \tag{4}$$

As shown in Table 4, For the resistance test, the $R_{i,grid}$ value for the grid test was 0.71; the $R_{i,time\ step}$ value for the time step was 0.65, confirming monotonic convergence; and the GCI_{21} value was 3.63%, while that for the time step was 2.13%, both of which were within acceptable limits. For the POW test, the $R_{i,grid}$ value for the grid test was 0.53, confirming monotonic convergence, and the GCI_{21} value was 2.07%, which is within acceptable limits. The medium grid condition and time step were selected considering the computational cost and GCI value.

Table 4. Calculation of discretization error for resistance and POW test.

	Resistance Test	POW Test (J = 0.35)	POW Test (J = 0.40)	POW Test (J = 0.45)
ε_{21}	0.711	4.785	2.066	1.530
ε_{32}	1.000	11.425	4.418	2.584
r	1.242	1.231	1.231	1.231
R	0.710	0.419	0.468	0.592
GCI^{21}_{fine}, %	3.626	2.747	1.691	2.451

2.5.2. Validation of Resistance Simulation

The calculation results were validated by comparing the total resistance coefficient with experiment data from towing tank model tests conducted at Pusan National University (PNU) using a model of the same scale. The dimensions of the tank are length, breadth, and depth of 100 m, 8.0 m, and 3.5 m, respectively. The resistance test and self-propulsion test were conducted under two conditions: loaded and ballast. The CFD simulation was compared at the loaded condition.

The comparison of the resistance performance in calm water without fin conditions indicated less than 2% error between the numerical calculations and the model tests, as shown in Table 5. Furthermore, resistance performance at various ship speeds for the with-fin condition was evaluated as shown in Table 6, confirming that the error in the total resistance coefficient from the model tests remained within 2%. Although absolute differences in sinkage and trim were observed, these variations were considered minor. Here, C_{TM}, τ, and σ represent the total resistance coefficient, trim, and sinkage, respectively.

Table 5. Comparison of resistance performance at design ship speeds (without flow control fin).

Case	C_{TM} ($\times 10^3$)	τ (deg)	σ/L_{PP} ($\times 10^2$)
Exp.	4.558	-	-
Present CFD	4.503	0.234	−0.217
Diff.	−1.21%		

The wake distribution influenced by the fin was analyzed (Figure 7a). The target ship exhibited a hook-shaped wake distribution; the flow velocity entering the top of the propeller increased, leading to a decrease in the nominal wake. These phenomena were further illustrated by the flow behavior around the fin, as shown in Figure 7b. The concentrated flow around the hook was dispersed near the fin, decreasing the low-speed region.

Table 6. Comparison of resistance performance at various ship speeds (with flow control fin).

Case	V_S (knots)	C_{TM} ($\times 10^3$)	τ (deg)	σ/L_{PP} ($\times 10^2$)
Exp.		4.207	0.158	−0.131
Present CFD	12.0	4.186	0.177	−0.180
Diff.		−0.49%	0.019	−0.049
Exp.		4.301	0.185	−0.157
Present CFD	13.0	4.311	0.220	−0.218
Diff.		0.23%	0.035	−0.061
Exp.		4.441	0.203	−0.175
Present CFD	13.5	4.451	0.225	−0.235
Diff.		0.23%	0.023	−0.061
Exp.		4.612	0.215	−0.190
Present CFD	14.0	4.709	0.237	−0.255
Diff.		1.90%	0.022	−0.065
Exp.		4.908	0.255	−0.220
Present CFD	15.0	4.970	0.269	−0.300
Diff.		1.26%	0.014	−0.080

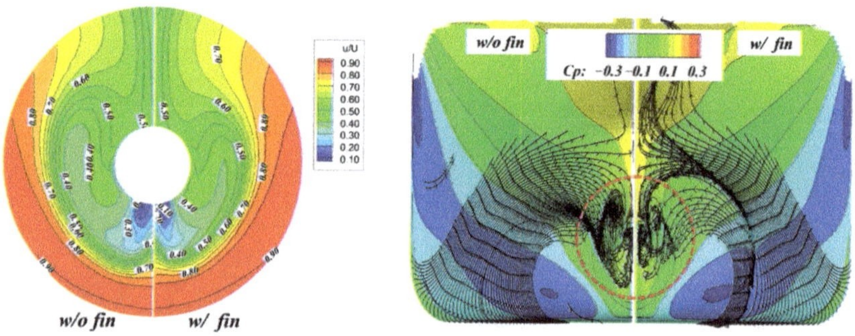

(a) (b)

Figure 7. Comparison of the flow distribution around the hull: (**a**) nominal wake; (**b**) streamline (black line) and pressure coefficient (red circle: propeller diameter).

2.5.3. Validation of Propeller Open-Water Simulation

The numerical results were validated by comparing the propeller performance factors, including thrust coefficient (K_T), torque coefficient ($10K_Q$), and propeller open-water efficiency (η_O), with experiment data, as shown in Table 7. The open-water curves were generated by calculating the flow velocity corresponding to each advance ratio (J) at a fixed revolution per second (RPS), as shown in Figure 8. K_T and $10K_Q$ exhibited decreasing trends with an increasing advance ratio, showing reasonable agreement between the experimental and CFD results. Near the design advance ratio, the numerical predictions revealed average deviations in thrust and torque of 2.0% and 1.7%, respectively, compared to the experimental data. These findings confirmed the accuracy and reliability of the numerical approach for predicting propeller performance.

Table 7. Propeller open-water test results of experiments and calculations.

J	Exp. K_T	Exp. $10K_Q$	Exp. η_O	Present CFD K_T	Diff.	Present CFD $10K_Q$	Diff.	Present CFD η_O	Diff.
0.10	0.2590	0.2505	0.1646	0.2565	−1.0%	0.2541	1.4%	0.1606	−2.4%
0.20	0.2211	0.2214	0.3179	0.2190	−0.9%	0.2245	1.4%	0.3106	−2.3%
0.30	0.1813	0.1902	0.4551	0.1801	−0.7%	0.1948	2.4%	0.4414	−3.0%
0.35	0.1612	0.1744	0.5150	0.1597	−1.0%	0.1787	2.5%	0.4976	−3.4%
0.40	0.1410	0.1584	0.5667	0.1387	−1.6%	0.1617	2.1%	0.5460	−3.6%
0.45	0.1205	0.1420	0.6078	0.1173	−2.6%	0.1436	1.1%	0.5850	−3.8%
0.50	0.0997	0.1250	0.6347	0.0957	−4.0%	0.1246	−0.3%	0.6112	−3.7%
0.55	0.0781	0.1068	0.6397	0.0735	−5.9%	0.1042	−2.4%	0.6175	−3.5%
0.60	0.0552	0.0868	0.6073	0.0506	−8.3%	0.0826	−4.8%	0.5852	−3.6%

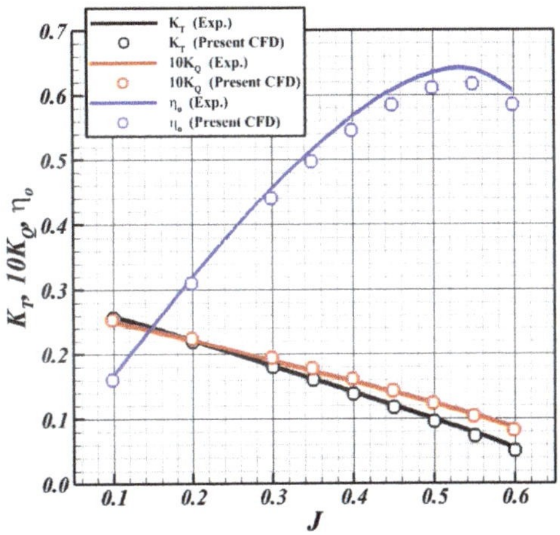

Figure 8. Comparison of the POW curves between experiment and simulation.

2.5.4. Validation of Self-Propulsion Simulation

The calculation results were validated by comparing the self-propulsion factors in calm water at the design velocity, as shown in Table 8. Instead of relying on the virtual disk theory commonly applied in numerical analysis, the propeller was modeled directly, and a self-propulsion simulation was conducted. The self-propulsion point was determined by balancing longitudinal forces, including the towing force, which is the skin friction correction force (*SFC*, F_D). The self-propulsion factors were estimated using the ITTC 1978 method [31].

Table 8. Self-propulsion factors in calm water.

	W_{TM}	W_{TS}	$1-t$	η_H	η_R	η_O	η_D	*RPM*	P_D (kW)
Exp.	-	-	-	1.128	1.002	0.569	0.643	204.5	2399
Present CFD	0.356	0.305	0.795	1.145	1.046	0.548	0.656	203.3	2344
Diff. (%)				1.5	4.4	−3.7	2.0	−0.6	−2.3

W_{TM}, W_{TS}, and *t* represent the model wake faction, full-scale wake fraction, and the thrust deduction, respectively.

Among the self-propulsion factors, the relative rotation efficiency (η_R) showed a larger difference, while the hull efficiency (η_H) differed by 1.5%. The quasi-propulsive efficiency (η_D), representing the ratio of delivered power to effective power (P_D), revealed a 2.0% difference, and the delivered power differed by 2.3%. These results showed reasonable agreement with the experimental data. Furthermore, trends consistent with the experimental results were confirmed through the numerical analysis validation conducted earlier, suggesting that the numerical scheme for this research has been established at a reliable level.

3. Methodology

3.1. Procedures for the Coanda Effect Performance Decision

Figure 9 shows the design point decision procedure for the Coanda propeller. The maximum efficiency was achieved when the torque became zero, following Newton's third law. In a Coanda hydrofoil, thrust increases, and torque decreases due to jet injection [24] because the propeller slits are oriented opposite to the direction of rotation. The jet injection reduces the torque acting on the rotating propeller. Hence, zero torque suggests no power is required to rotate the propeller. Consequently, maximum efficiency is achieved when the torque reaches zero, a condition referred to as the Coanda design point.

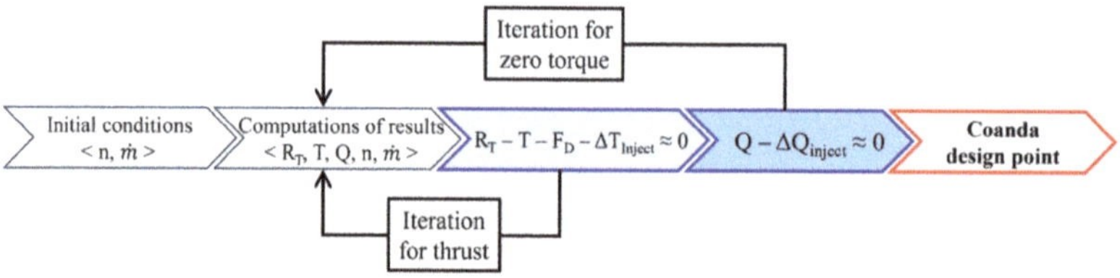

Figure 9. Coanda design point decision procedure.

The rotational speed must be adjusted because the amount of jet injections affects the torque and thrust. Accordingly, iterative calculations were carried out to maintain a constant thrust while ensuring the condition where the torque becomes zero is satisfied.

In general, a propulsion test involves installing a propeller on a model ship and evaluating the propulsion performance based on the thrust generated by the propeller. During this test, data are collected under various combinations of rotational speeds and velocity conditions to evaluate the efficiency and performance of the propeller. A self-propulsion test was conducted in a similar manner to the model test and was carried out at a corresponding speed that satisfies the similarity in the Froude number between the model and the full-scale ship. On the other hand, the Reynolds number similarity cannot be satisfied because of the scale difference between the model and the full-scale ship. Accordingly, a towing force was applied to the model to replicate the actual conditions of the full-scale ship. The conditions of Equation (5) were satisfied at this self-propulsion point.

When a Coanda propeller is used, the jet injection amount must be adjusted to achieve zero torque. Nevertheless, this adjustment directly affects the thrust, requiring the fluctuating thrust to be considered when determining the self-propulsion point (Equation (6)). Unlike conventional methods, self-propulsion tests with a Coanda propeller necessitate an iterative optimization process that accounts for the interaction between the jet injection amount and thrust. In this study, the point that satisfies these conditions was defined as the self-propulsion point of the Coanda propeller.

$$R_T - T - F_D = 0 \tag{5}$$

$$R_T - T - F_D - \Delta T_{inject} = 0 \tag{6}$$

where R_T, T, F_D, and ΔT_{inject} represent the total resistance, propeller thrust, towing force, and the thrust generated by the Coanda effect, respectively.

3.2. Procedures for Pump Performance

The Coanda propeller requires the inflow of fluid into its interior, necessitating using a pump to handle the intake of separate fluids. In this study, a centrifugal pump was chosen because of its widespread application in converting mechanical energy from an external source (e.g., an electric motor) into the kinetic energy of the fluid, facilitating efficient fluid transfer. The primary objective of the pump is to achieve the desired flow rate at the specified head. The power required by the pump is categorized into water horsepower (WHP, P_W), brake horsepower (BHP, P_B), and motor power (i.e., total required power, P_M), as shown in Figure 10.

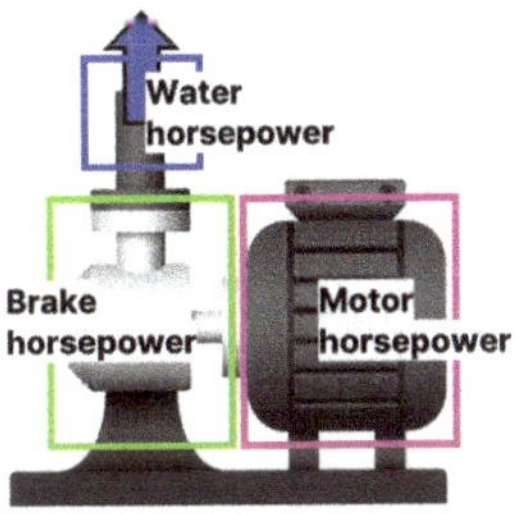

Figure 10. Conceptual diagram of pump power.

First, water horsepower refers to the power imparted to the fluid by the pump, determined by the total head and flow rate. The total head is the sum of the static head, velocity head, pressure head, and friction loss head. Second, brake horsepower refers to the power adjusted for the losses incurred during power transmission from the pump to the fluid, factoring in the pump efficiency. The pump efficiency is determined from a performance curve that relates the flow rate to specific impeller rotation speeds and size conditions. Finally, motor power is the power required to operate a motor, accounting for the brake horsepower and the transmission coefficient (i.e., margin rate). In this study, the power required to drive the pump was calculated using the formula provided in the technical guidelines of the Korea Occupational Safety and Health Agency (KOSHA), as expressed in Equation (7).

$$
\begin{gathered}
P_W = \gamma \times Q \times H_{total} = \dot{m} \times g \times H_{total}\,, \\
P_B = P_W \div \eta_P\,, \\
P_M = P_B \div \eta_M
\end{gathered}
\tag{7}
$$

where γ, Q, H_{total}, $\dot{m}$, g, η_P, and η_M represent the specific weight, volumetric flow rate, total head, mass flow rate, gravitational acceleration, pump efficiency, and motor efficiency including a margin, respectively.

4. Results of the Simulation

Three types of propellers were attached to the stern of a ship, and simulations were performed to compare the self-propulsion factors at the self-propulsion point on a model scale. In addition, this study explored the performance and practical application of propellers using the Coanda effect under stern wake conditions. The self-propulsion point on the model scale refers to the condition where the model ship achieves a specific forward speed under its propulsion. At this point, the thrust generated by the propeller equals the hull resistance, eliminating the need for additional towing force. Additional power components, such as pump power, cannot be extrapolated reliably to full-scale conditions because of the influence of the scale ratio. Consequently, this study compared the performance at the model-scale self-propulsion point. Future full-scale analyses will focus on evaluating the applicability of Coanda propellers to full-size ships and assessing their economic viability.

The water horsepower required for injection was incorporated into the propeller efficiency calculation, as shown in Equation (8) [23,24]. In this study, the pump efficiency and the motor margin were set to 50% and 25%, respectively. The relative rotational efficiency of the Coanda propeller was set to 1.0 because no torque acts on the propeller.

$$\eta_O = \frac{TV_A}{2\pi nQ + P_W} \tag{8}$$

$$\eta_D = \eta_H \times \eta_O \times \eta_R \times \eta_P \times \eta_M \tag{9}$$

In this study, the efficiency was evaluated from two perspectives. The first perspective focused on the propeller and ship efficiency, which followed a conventional definition used in existing propulsion systems. The second perspective considered the efficiency associated with the Coanda effect, which accounted for additional thrust not included in the conventional definition. This efficiency is expressed as the product of hull efficiency (η_H), propeller efficiency (η_O), relative rotational efficiency (η_R), pump efficiency (η_P), and motor efficiency (η_M) with margin rate, as expressed in Equation (9). Table 9 lists the self-propulsion simulations for different propellers, which compares the self-propulsion factors.

Table 9. Comparison of self-propulsion performance at the design ship speeds (without flow control fin).

Case	Original Prop.	Normal Prop.	Coanda Prop.
RPS_M	15.865	16.724	15.012
K_{TM}	0.163	0.145	0.176
Q_{TM} (Nm)	0.833	0.859	-
P_W (W)	-	-	24.7
w_{TM}	0.372	0.326	0.280
t_M	0.159	0.150	0.128
η_H	1.340	1.261	1.211
η_R	1.035	1.031	1.000
η_O	0.506	0.494	1.881
η_P	-	-	0.500
η_M	-	-	0.667
η_D	0.701	0.643	0.759
P_E (W)	56.3	56.3	56.3
P_D (W)	80.2	87.6	74.1
Diff. of P_D (%)	-	9.0	−7.8

Additional thrust is generated because of the Coanda effect, allowing the Coanda propeller to produce similar thrust to other propellers even at lower rotation speeds. As a result, higher advance ratios were observed during operation. The lowest thrust deduction factor was observed in the Coanda propeller, suggesting that relatively lower resistance was experienced at the self-propulsion point than other propellers. Figure 11 compares the pressure distributions around the stern between the Coanda propeller and normal propeller. A wider high-pressure region near the stern was observed with the Coanda propeller compared to the normal propeller. This high-pressure region was attributed to the reduction in resistance achieved through the pressure recovery effect in the hull.

Figure 12 shows the wake vortices of the three propellers at Q = 1000 s^{-2}. A complex vortex system was observed in the wake, which was induced by the tips and the hub. The Coanda propeller generated secondary vortices along the propeller plane as the jet was injected. Figure 13 shows the instantaneous vorticity magnitude in the XZ plane. Relatively organized vortex patterns are shown in the downstream region for the original and normal propellers, with the vortices originating from the blade tips and the hub. In contrast, irregular vortex shedding is observed in the Coanda propeller because of interactions between the jet and the tip and hub vortices. The distance at which the wake transitions to turbulence is reduced by the jet injection. Furthermore, the dissipation of vorticity over shorter distances is identified, showing the enhanced vortex dispersion characteristics caused by the Coanda effect.

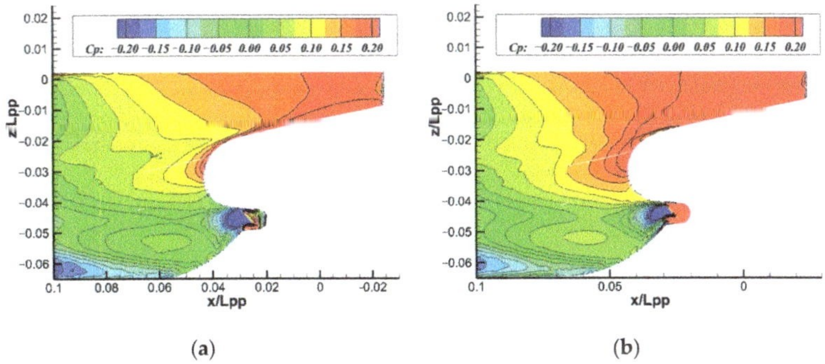

Figure 11. Comparison of the pressure distribution around the stern. (**a**) nominal propeller; (**b**) Coanda propeller.

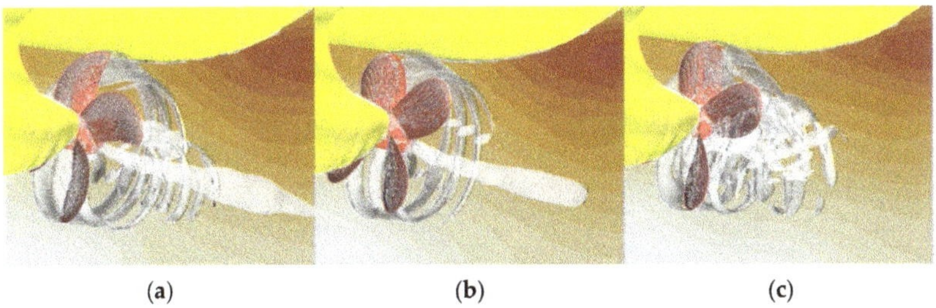

Figure 12. Vortices visualized based on the Q-criterion at the stern: (**a**) original propeller; (**b**) normal propeller; (**c**) Coanda propeller.

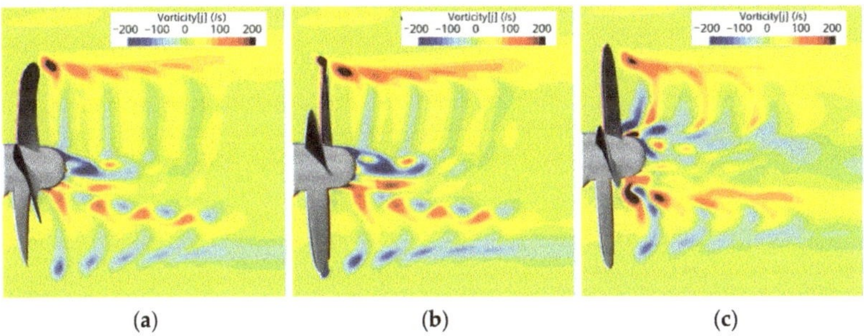

Figure 13. Vorticity intensity in the XZ plane around the propeller: (**a**) original propeller; (**b**) normal propeller; (**c**) Coanda propeller.

The horsepower required by the three propellers was compared. The Coanda propeller required approximately 7.8% less power than the original propeller. These findings suggest that a reduction in required horsepower can be achieved using the Coanda propeller, showing a decrease in the energy or power needed for a ship to achieve a specific speed.

As a result, the Coanda propeller improves ship efficiency, showing its practical feasibility behind the ship. The findings, derived through horsepower comparisons, confirmed that the Coanda effect enhances performance compared to conventional propellers. These results provide essential baseline data for evaluating the applicability of Coanda effect-based propellers and are expected to be validated through full-scale analyses in future studies.

5. Conclusions

This study numerically investigated the self-propulsion performance of a propeller using the Coanda effect. A propeller incorporating a jet injection slit to implement the Coanda effect was designed, and its performance was compared with the original propeller. Validation was conducted by comparing the numerical results with the experimental data, and the results were verified using the GCI test. In addition, the self-propulsion performance at the model scale was analyzed, and the self-propulsion factors were compared to evaluate the performance of the Coanda propeller.

- The Coanda propeller showed significant improvements in the self-propulsion performance compared to the normal and original propellers. The self-propulsion analysis indicated that the required power of the Coanda propeller was reduced by approximately 7.8% compared to the original propeller. These results suggest that additional force is generated on the suction side because of jet injection, enabling efficient propulsion even at lower rotational speeds.
- The numerical scheme used in this study was based on the URANS equations and showed a correlation with the model experimental results. This approach effectively reproduced key flow phenomena and accurately modeled the flow characteristics of jet injection because of the Coanda effect, allowing precise thrust and torque calculations.
- The numerical results confirmed that the Coanda effect positively influenced the flow characteristics by stabilizing the wake and enhancing pressure recovery. In the wake region, the interactions between tip vortices and the jet flow were observed, promoting vortex dissipation, reducing resistance, and improving propulsion efficiency. Furthermore, pressure fluctuations showed that the Coanda propeller exhibited lower pressure variations than normal and original propellers, indicating its potential for cavitation.

This study identified the self-propulsion performance and flow characteristics of the Coanda propeller on the model scale, providing foundational data for extending the analysis to full-scale conditions. Future research will focus on incorporating Reynolds number effects and free-surface interactions in full-scale studies, as well as refining propeller geometry design to determine the practical applicability of this technology.

Author Contributions: Conceptualization, Data curation, Formal analysis, Investigation, Methodology, Validation, Visualization, Writing—original draft, review, and editing, J.-H.L.; Conceptualization, Data curation, Methodology, Funding acquisition, Investigation, Resources, Supervision, Writing—review and editing, K.-J.P.; Conceptualization, Investigation, Visualization, Writing—review and editing, D.-H.O. All authors have read and agreed to the published version of the manuscript.

Funding: This work was supported by the National Research Foundation of Korea (NRF) grant (No. 2022R1F1A1074865) funded by the Ministry of Science and ICT, Republic of Korea, and the Korea Institute of Marine Science and Technology Promotion (KIMST) grant (20220210) funded by the Ministry of Oceans and Fisheries, Republic of Korea.

Institutional Review Board Statement: Not applicable.

Informed Consent Statement: Not applicable.

Data Availability Statement: Data are contained within the article.

Conflicts of Interest: The authors declare no conflicts of interest.

References

1. Lee, J.Y. International Maritime Organization (IMO) 82nd Marine Environment Protection Committee (MEPC). *J. Soc. Nav. Archit. Korea* **2024**, *61*, 12–21.
2. Kim, J.H.; Choi, J.H. A study on impact and countermeasures of marine fuels in the FuelEU Maritime regulation. *J. Soc. Nav. Archit. Korea* **2024**, *61*, 88–97. [CrossRef]
3. Su, Y.M.; Lin, J.F.; Zhao, D.G.; Guo, C.Y.; Guo, H. Influence of a pre-swirl stator and rudder bulb system on the propulsion performance of a large-scale ship model. *Ocean Eng.* **2020**, *218*, 108189. [CrossRef]
4. Koushan, K.; Krasilnikov, V.; Nataletti, M.; Sileo, L.; Spence, S. Experimental and numerical study of pre-swirl stators (PSS). *J. Mar. Sci. Eng.* **2020**, *8*, 47. [CrossRef]
5. Kang, J.G.; Kim, M.C.; Shin, I.R.; Jin, W.S. Feasibility study on effect of structural flexibility of asymmetric pre-swirl stator on propulsion performance for kriso container ship (Kcs). *Brodogradnja* **2021**, *72*, 103–119. [CrossRef]
6. Gaggero, S. Design of PBCF energy saving devices using optimization strategies: A step towards a complete viscous design approach. *Ocean Eng.* **2018**, *159*, 517–538. [CrossRef]
7. Ouchi, K.; Ogura, M.; Kono, Y.; Orito, H.; Shiotsu, T.; Tamashima, M.; Koizuka, H. A research and development of PBCF (propeller boss cap fins): Improvement of flow from propeller boss. *J. Soc. Nav. Archit. Jpn.* **1988**, *163*, 66–78. [CrossRef]
8. Ouchi, K.; Tamashima, M.; Kawasaki, T.; Koizuka, H. A research and development of PBCF (propeller boss cap fins), 2nd report: Study on propeller slipstream and actual ship performance. *J. Soc. Nav. Archit. Jpn.* **1989**, *165*, 43–53. [CrossRef]
9. Xu, K.; Su, X.; Bensow, R.; Krajnovic, S. Drag reduction of ship airflow using steady Coanda effect. *Ocean Eng.* **2022**, *266*, 113051. [CrossRef]
10. Pfingsten, K.C.; Radespiel, R. Numerical simulation of a wing with a gapless high-lift system using circulation control. In *New Results in Numerical and Experimental Fluid Mechanics VI: Contributions to the 15th STAB/DGLR Symposium*; Springer: Berlin/Heidelberg, Germany, 2007.
11. Burnazzi, M.; Radespiel, R. Design and analysis of a droop nose for Coanda flap applications. *J. Aircr.* **2014**, *51*, 1567–1579. [CrossRef]
12. Lee, H.; Han, S.; Lee, H.; Jeon, J.; Lee, C.; Kim, Y.B.; Choi, H.R. Design optimization, modeling, and control of unmanned aerial vehicle lifted by Coandă effect. *IEEE/ASME Trans. Mechatron.* **2017**, *22*, 1327–1336. [CrossRef]
13. Jun, H.C.; Kim, S.H.; Kim, H.J.; Park, H.P. An experimental study on the effect of increasing rudder force on the turning ability of a short sea shipping ship. *J. Korean Soc. Mar. Environ. Saf.* **2012**, *18*, 591–596. [CrossRef]
14. Maltsev, L.I.; Dimitrov, V.D.; Milanov, E.M.; Zapryagaev, I.I.; Timoshevskiy, M.V.; Pervunin, K.S. Jet control of flow separation on hydrofoils: Performance evaluation based on force and torque measurements. *J. Eng. Thermophys.* **2020**, *29*, 424–442. [CrossRef]
15. Oh, J.K.; Noh, J.K. On the lift enhancement technique of the trailing edge rotor of two-dimensional hydrofoil. *J. Korean Soc. Mar. Environ. Saf.* **2015**, *21*, 200–206. [CrossRef]
16. Oh, J.K.; Kim, H.C. A study on the high lifting device equipped with the trailing edge rotor for the enhancement of circulation control. *J. Soc. Nav. Archit. Korea* **2010**, *47*, 533–542.
17. Pant, C.S.; Frankel, S.H. Interaction between surface blowing and re-entrant jet in active control of hydrofoil cavitation. *Ocean Eng.* **2021**, *242*, 110087. [CrossRef]
18. Wang, W.; Tang, T.; Zhang, Q.D.; Wang, X.F.; An, Z.Y.; Tong, T.H.; Li, Z.J. Effect of water injection on the cavitation control: Experiments on a NACA66 (MOD) hydrofoil. *Acta Mech. Sin.* **2020**, *36*, 999–1017. [CrossRef]
19. Chowdhury, N.E.; Hossain, M.F.; Karim, M.M.; Islam, M.R.; Bhuiyan, A.A. Numerical investigation for mitigation of cavitation in high-speed marine propeller using mass injection approach. *Iran. J. Sci. Technol. Trans. Mech. Eng.* **2023**, *47*, 1693–1709. [CrossRef]
20. Yang, C.; Yao, H.D.; Sun, C.; Guo, C.; Wang, C.; Ren, W. Injection of micro jets to improve hydrodynamic performance of a ducted propeller. *Ocean Eng.* **2024**, *309*, 118502. [CrossRef]
21. Eom, M.J.; Paik, K.J.; Lee, J.H. Numerical study on the lift-drag and cavitation performances of a two-dimensional hydrofoil using the Coanda effect. *J. Adv. Mar. Eng. Technol.* **2020**, *44*, 457–466. [CrossRef]
22. Eom, M.J.; Paik, K.J.; Lee, J.H.; Kang, S.M.; Kim, D.Y. A study on the performance variation of a three-dimensional hydrofoil using jet flow. *J. Ocean Eng. Technol.* **2021**, *35*, 24–37. [CrossRef]
23. Lee, J.H.; Paik, K.J.; Lee, S.H.; Kim, G.H.; Cho, J.H. Experimental and numerical study on the performance change of a simple propeller shape using the Coanda effect. *Appl. Sci.* **2021**, *11*, 4112. [CrossRef]
24. Lee, J.H.; Lee, J.H.; Kim, M.M.; Oh, D.; Paik, K.J. A numerical study on the optimization of the slit shape of a jet injection propeller. *Int. J. Nav. Archit. Ocean Eng.* **2024**, *16*, 100578. [CrossRef]
25. Menter, F.R. Two-equation eddy-viscosity turbulence models for engineering applications. *AIAA J.* **1994**, *32*, 1598–1605. [CrossRef]
26. Lee, S.H.; Paik, K.J.; Cho, J.H.; Kim, G.H.; Kim, H.S.; Kim, S.H. A Numerical Study on Self-Propulsion Performance in Regular Waves and Power Prediction in Irregular Waves. *Int. J. Nav. Archit. Ocean Eng.* **2022**, *14*, 100454. [CrossRef]

27. Grlj, C.G.; Degiuli, N.; Martić, I. Experimental and numerical assessment of the effect of speed and loading conditions on the nominal wake of a containership. *Brodogradnja* **2024**, *75*, 75405. [CrossRef]
28. Simens. *STAR-CCM+ User Guide, Version 16.06*; Simens: Munich, Germany, 2021.
29. Celik, I.B.; Ghia, U.; Roache, P.J.; Freitas, C.J.; Coleman, H.; Raad, P.E. Procedure for Estimation and Reporting of Uncertainty Due to Discretization in CFD Applications. *J. Fluids Eng.* **2008**, *130*, 078001.
30. ITTC. Uncertainty Analysis in CFD Verification and Validation Methodology and Procedures. *ITTC-Rcomm. Proced.* 7.5-03-01-01.4. 2021. Available online: https://www.ittc.info/media/11950/75-03-01-01.pdf (accessed on 23 February 2025).
31. ITTC. 1978 ITTC Performance Prediction Method. *ITTC-Rcomm. Proced.* 7.5-02-03-01.4. 2018. Available online: https://www.wartsila.com/encyclopedia/term/ittc-performance-prediction-method (accessed on 23 February 2025).

MDPI AG
Grosspeteranlage 5
4052 Basel
Switzerland
Tel.: +41 61 683 77 34

Journal of Marine Science and Engineering Editorial Office
E-mail: jmse@mdpi.com
www.mdpi.com/journal/jmse

MDPI

www.ingramcontent.com/pod-product-compliance
Lightning Source LLC
Chambersburg PA
CBHW061551210926
51850CB00009B/2071
9783725878819